Palgrave Studies in Environmental Policy and Regulation

Series editor
Justin Taberham
London, UK

The global environment sector is growing rapidly, as is the scale of the issues that face the environment itself. The global population is estimated to exceed 9 billion by 2050. New patterns of consumption threaten natural resources, food and energy security and cause pollution and climate change. Policy makers and investors are responding to this in terms of supporting green technology as well as developing diverse regulatory and policy measures which move society in a more 'sustainable' direction. More recently, there have been moves to integrate environmental policy into general policy areas rather than having separate environmental policy. This approach is called Environmental Policy Integration (EPI). The series will focus primarily on summarising present and emerging policy and regulation in an integrated way with a focus on interdisciplinary approaches, where it will fill a current gap in the literature.

More information about this series at
http://www.palgrave.com/series/15053

Jon Foreman
Editor

Developments in Environmental Regulation

Risk based regulation in the UK and Europe

palgrave
macmillan

Editor
Jon Foreman
Bristol, UK

Palgrave Studies in Environmental Policy and Regulation
ISBN 978-3-319-61936-1 ISBN 978-3-319-61937-8 (eBook)
DOI 10.1007/978-3-319-61937-8

Library of Congress Control Number: 2017949908

Cover illustration: Sara Berdon/Alamy Stock Photo

Printed on acid-free paper

This Palgrave Macmillan imprint is published by Springer Nature
The registered company is Springer International Publishing AG
The registered company address is: Gewerbestrasse 11, 6330 Cham, Switzerland

WaterAid UK

All proceeds from the sale of the book will be donated to WaterAid UK, which is working in countries across the world to transform millions of lives every year by improving access to safe water, toilets and hygiene. For further information visit www.wateraid.org/uk

Foreword

Regulation is one of the ways society ensures that businesses and individuals carry out activities in ways that minimise adverse effects on people, wildlife and the environment. It is often used when the impacts of external factors are not fully reflected in the economic costs of an activity. Regulation establishes minimum operating requirements and provides confidence to the public that there are appropriate checks on what businesses and individuals are doing.

A substantial body of environmental regulation has been developed and implemented in many countries. This has resulted in significant benefits for people, wildlife and the environment, as levels of air and water pollution have been reduced, waste and natural resources better managed and biodiversity provided greater protection. However, over time, there can be an accumulation of policy and regulatory instruments. This can increase the complexity of the regulatory landscape and result in a lack of clarity in what businesses and individuals have to do to comply.

A periodic review of regulations is necessary to ensure that the various requirements are coherent and consistent and that they are compatible with effective and efficient regulatory approaches that minimise administrative costs and bureaucracy whilst maintaining the necessary protections for people, wildlife and the environment.

In recent years, many governments have sought to reduce costs to businesses and facilitate economic growth. A concept of better regulation has been widely promoted within many jurisdictions that seeks to achieve policy and regulatory outcomes at less cost for all involved in the delivery of the regulatory requirements. One aspect of such programmes has been a review of the cost regulations placed on businesses to see if these can be reduced. Governments and their regulators have also sought to improve the implementation of regulation through the use of risk-based approaches, targeting regulatory effort towards the greatest risks. I believe that the UK has been at the forefront of the development and implementation of risk-based approaches.

There will be both challenges and opportunities in terms of environmental protection and improvement associated with Brexit. It is vital that necessary regulatory requirements and controls are maintained for the benefit of people and wildlife. However, the opportunity to reduce unnecessary costs and bureaucracy should also be taken. The focus of a better regulation agenda has to be on reducing costs and bureaucracy whilst maintaining the necessary protection for people, wildlife and the environment.

Professor of Environmental Assessment
Cranfield University,
Cranfield, UK

Paul Leinster

Preface

'Developments in environmental regulation' aims to shed light on the mystery of environmental regulatory practice in the UK and the European Union (EU). It is not a book so much about the details of legislation but rather it examines how regulation has developed and has been implemented and therefore how it affects business. It provides a snapshot to enable practitioners who may be policy-makers, regulators, environmental managers in the industry or students about to embark on a career in this field to gain an understanding of the approach of regulatory organisations. It is hoped that the book will enable regulatory practice to be not only better understood but encourage a collaborative approach for the benefit of people and the environment which we all share.

We live in a time of momentous geo-political change and following the UK's move to commence withdrawal from the EU, environmental regulation is facing a significant period of adjustment. Environmental regulation in the UK is overwhelmingly shaped by the legislation and policy of the EU, as it has been for over 40 years. And while 'Brexit' may bring some opportunities for simplification, the overall picture is that UK environmental policy and regulation will continue to be intricately dependent on the policy direction of the EU, our largest trading partner. At various points, 'Developments in environmental regulation' will look to the future and examine the issues facing the UK as it gears up for 'Brexit' and a future on the open seas of free trade.

The views presented in this book are those of the authors and no attribution is intended. Any comments or queries on any aspect of this book can be directed to any of the authors by email or via linkedin.

www.jonforeman.co.uk/

Jon Foreman Associates
Bristol, UK

Jon Foreman

Acknowledgements

The book has been written by a group of highly experienced regulatory specialists who have worked for government, governmental organisations and industry in the UK, Europe and worldwide. I am extremely grateful to them for their contributions and for the enthusiasm they have shown in helping to compile this collection of essays. I would also like to thank Paul Leinster, Professor of Environmental Assessment at Cranfield University, and former Chief Executive of the Environment Agency, England and Wales, who has kindly written the foreword to the book.

Authors acknowledge permission to use the following material:

The Aldersgate group (2011). *Dealing with Deficits. Best value Regulation to reduce our environmental and financial debts*

Baldrige Performance Excellence Program. (2017). *Baldrige Excellence Framework—graphics*

Chartered Quality Institute (CQI) (2016). Integrated Management Special Interest Group, *illustrations*

Defra's Smarter Environmental Regulation Review

Engineering Employers' Federation. (2015). *Green Tape: Manufacturers' Views of Progress*

IEMA (2016) Environmentalist Journal, October 2016 *Emisoft Statoil case study*

IMPEL (EU Network for the Implementation and Enforcement of Environmental Law) projects, (2011–16): *Experience of Derogations from IED BAT-AELs; Supporting IED Implementation; Implementation of the iDepend decision support tool; Choosing Appropriate Interventions, Phase 3; Compliance assurance through company compliance management systems.*

Institute for European Environmental Policy. (2016). *Make it Work*

Oliver, R. (2014). 'Smarter and wiser? An update on Defras Smarter Environmental Regulation Review

Oliver, R. (2015). *Environmental regulation—stripping back bureaucracy or protection?*

Tombs, S. (2016). *Better Regulation for Whom. Briefing 14 April 2016.* London: Centre for Crime and Justice Studies.

UK Environmental Law Association. (2012). *The State of UK Environmental Legislation in 2011–12: Is there a case for legislative reform?*

Viridor, (2011). *Openspace Web Portal developed with Environment Agency.*

We are also indebted to the many individual researchers and organisations whose work is widely quoted throughout the book, including:

- The Ends report
- The European Commission, European Parliament and European Union
- UK Environmental Regulators: Environment Agency, Scottish Environment Protection Agency and Northern Ireland Environment Agency, Natural Resources Wales
- UK Government Departments and Devolved Administrations, particularly Defra, BEIS, Parliamentary Committees, Hansard publications

Contents

List of Figures

List of Tables

1

Introduction to Environmental Regulation and Practice in the UK and Europe

Martin Bigg

1.1 Introduction

Environmental regulation has an important role in achieving a clean, healthy and sustainable environment. It protects people, animals and plants from damaging actions and harmful chemicals. It ensures that we use our resources including energy, and manage our wastes, in a more sustainable way. Environmental regulation can set the minimum requirements for an activity or impact and can create a level playing field between operators, sectors and countries. It can encourage innovation and new technologies as well as curtail unacceptable behaviours.

Environmental regulation can take many forms, from the enforcement of a numerical limit or standard (a driving comparison is the speed limit) to operational controls (the tachograph in the cab), to voluntary initiatives (fitting a speed limiter) and agreements (buying a different vehicle). The success of these actions is usually only as good as the quality/competency of the operator (or driver) and the enforcement authority. If the amount

M. Bigg (✉)
University of the West of England (UWE), Bristol, UK

J. Foreman (ed.), *Developments in Environmental Regulation*, Palgrave Studies in Environmental Policy and Regulation, DOI 10.1007/978-3-319-61937-8_1

of monitoring or the number of enforcers is reduced, non-compliance tends to increase. In any activity, while the majority will comply because of the regulations, there is always a small number who will try to avoid the regulations. It is important that the regulatory tools are kept under review to ensure that they are achieving their objectives in the most cost-effective way and keeping ahead of the changing world.

In an uncertain world, one certainty is change. Our environment is continually changing because of natural processes and human interventions. Over the past 150 years we have seen radical changes to industrial and business practices, materials used, wastes and pollutants produced. We have improved our understanding of the causes, effects and consequences of environmental damage but we still face surprises and are slow to adjust to the threats, ranging from climate change to managing our waste. Our environmental regulation and practices in the UK and Europe have often struggled to keep up with all these changes, occasionally making significant steps but usually being slow to evolve, for example, in responding to the need to tackle climate change. The public, governments and businesses can have differing perceptions and priorities, but when they converge, commitment and change can be quicker, for example, the introduction of a carrier bag charge (UK Parliament 2015).

The vote to leave the European Union (EU) has the potential to have a dramatic impact on environmental regulation. We will need to ensure the continuation of the hard-won Europe-wide baseline for environmental protection on the likes of waste, habitats or air quality and maintain alliances with European regulators, operators and policy makers. We will also need to ensure that environmental protection, climate change and sustainability measures are not lost or diluted in the rush to negotiate new trade agreements around the globe.

This book focuses on the environmental regulation of businesses, more than on individuals. This chapter provides a snapshot of where we are now, the lessons and experiences from the past and, most importantly, pointers as to the future of environmental regulation and practice. Subsequent chapters address the balancing of the benefits and burden of environmental regulations and the drive for deregulation, better regulation and self-regulation.

1.2 Overview of the Regulatory Landscape

We all need regulations, whether they are controls on the manufacturing of toys to avoid injury to children or on car drivers to avoid accidents. We all tend to push at the requirements of these regulations until they are enforced. In 2015, over 1.2 million UK drivers attended speed awareness courses for having exceeded the speed limit (National Driver Offender Retraining Scheme 2016), and according to the Royal Automobile Club (RAC) foundation, the transport policy and research organisation, three-quarters of a million fixed penalty notices were issued for speeding in 2014. Despite knowing the reasons for and benefits of road safety regulation we continue to push the boundaries.

National UK laws were developed to address issues across the country and more recent national legislation has been in response to global or regional international agreements. The EU has increasingly been the primary source of environmental regulation legislation with member states transposing it into national law.

Under the Ordinary Legislative Procedure (European Council 2016), the Commission submits a proposal to the European Parliament (EP) and Council of Ministers (Council) who either approve or amend it. If the EP and Council cannot reach an agreement on the proposed amendments, both can amend the proposal a second time. If they still cannot reach an agreement they enter into negotiations. Following the conclusion of these negotiations, both institutions can vote either in favour or against the proposal.

There have been regular complaints of "gold plating" of EU legislation made in some countries while other countries have been accused of a lack of legislation or poor enforcement (Davidson 2006; Gerda Falkner et al. 2004). In the UK and across Europe, there is a drive for deregulation, modern regulation or better regulation (European Commission 2016). In parallel, there has been a significant drive to cut costs and expenditure. This has resulted in the development of new approaches to the achievement of environmental outcomes, the reduction in some levels of inspection and enforcement and the loss of significant amounts of guidance. It is debatable whether we have yet seen all the benefits and savings to regulated activities, the public or the environment (Hjerp et al. 2010).

There is significant confusion and uncertainty in the public and business mind as to what is "environmental regulation". It includes:

- EU directives (which require national states to implement in their own legislation (transposition))
- EU regulations (which apply directly to a member state without the need for national legislation)
- EU decisions (which have direct effect but only relating to a specific person or entity)
- UK primary and secondary legislation
- local authority laws

Many also use the words "environmental regulation" to embrace the guidance and agreements between regulators, industry, industry bodies and third parties. Some regulations contain clearly defined limits with details on how they are applied. Others contain requirements such as the use of "best available techniques (BAT)" in the EU industrial emissions directive (IED), which require greater clarification. The EU's BAT reference documents (BREF) and UK guidance give BAT-associated emission levels (BAT-AEL). Subsequent sections provide further details on this directive.

Because of this complex mix of regulation and guidance, much of the enforcement and enforceability of environmental regulation is dependent on a common understanding between a company and their regulator. In many cases, this is dependent on the understanding of the individual officer and their business/industrial counterpart. As a result, there can be resistance by an industry or an individual company to changes in their respective regulator and charges of regulatory capture by third parties. The regulatory bodies address this issue by ensuring that there is movement of regulatory staff, the use of local and national regulatory teams and robust procedures for monitoring, inspection and enforcement.

With increasing restrictions on resources and the very diverse range of activities and environmental impacts regulated, it is increasingly difficult to achieve site-specific expertise and flexible regulation. For large businesses and across industry sectors, a dedicated regulator account manager may have oversight of the working relationship. For industrial processes, a cost-benefit analysis is used to determine what the regulator considers BAT. If the costs are disproportionate to environmental benefits, the IED allows authorities to disregard BAT-AELs when setting permit conditions.

The referendum vote by 72.2% of the UK population on 23 June 2016 with 51.9% voting to leave the EU and 48.1% voting to remain is likely to have a significant effect on UK environmental regulation. Although environmental issues were not a significant debating subject during the build-up to the referendum, environmental groups were concerned that some high-profile supporters of the campaign to leave were sceptical about the impact of climate change and the need to address it. Environmental groups were also concerned that the strong campaign to reduce the amount of regulation by leaving the EU could result in the loss of important environmental protection regulations (Ends 2016).

Many UK laws derive from or are associated with EU policy and legislation. Most of these laws are directly linked with directives, for example, the Waste Framework Directive and the Birds and Habitats Directives. When the European Communities Act 1972 is repealed, measures are needed to retain the requirements of the directives in UK Law. The legislation could be unpicked or amended at leisure. If the UK chooses to or is able to retain some engagement with the single market, then it is very likely that it will need to continue to comply with EU environmental and product legislation.

A strong feature of the campaign to leave the EU was the criticism of the bureaucracy and red tape from allegedly undemocratic and unelected bureaucrats in Brussels. However, specifically, criticism of EU environment legislation up to the referendum was small, focusing on a few limited and sometimes mischievous interpretations of the legislation such as the controls on the power consumption of vacuum cleaners or the curvature of bananas (Perring 2016). As politicians who campaigned to leave the EU now feel more empowered, it is more likely that a wider range of EU-derived legislations will come under attack, particularly where there is pressure for new energy generation or transport infrastructure to boost growth or investor confidence. The UK was a significant player in the negotiations on climate change and the Paris Agreement and pushed for ambitious climate change objectives and energy management measures. Even if the UK continues its commitments to the EU emissions trading scheme, it will have little influence in its future development.

The UK government under Prime Minister David Cameron made significant commitments to addressing climate change through the earlier government's Climate Change Act 2008 and subsequent carbon budgets but the future of these commitments is uncertain. UK air quality should be protected by the 2008 Ambient Air Quality Directive and in particular the nitrogen oxides limits. However, the UK continues to be in breach, despite court action including the UK Supreme Court. With nominally two years to leave the EU, it is likely that any action will be pushed beyond this time scale. Civil servants and ministers are significantly distracted and resources too stretched for the development, revision or implementation of significant environmental legislation over at least the next two years.

1.3 The Development and Evolution of Industrial Regulation in the UK

Environment law in the UK derives from common law in which important principles were and still are established by judges deciding cases rather than through legislation. People used common law to protect themselves against pollution of the air, land or water and the damage it did to health and property. Action could be taken against a private nuisance to the value or enjoyment of property, which may have been accidental or intentional. Public nuisance could be used to tackle an act that caused damage to public health.

In the nineteenth century, serious pollution of air and water was tackled by measures to prevent "nuisance". The Nuisances Removal and Diseases Prevention Act 1848, gave powers to local bodies to take action to protect health including inspecting drains. The Metropolitan Commission of Sewers Act 1848, banned the use of cesspits in the cities of London and Westminster and resulted in waste being dumped in the Thames.

Foul smelling air or poisonous vapour (miasma) was blamed for the spread of disease. It resulted in the first Public Health Act in 1848, setting up of a Board of Health and giving towns the right to appoint a Medical Officer of Health.

In the UK and across Europe, early environmental regulation targeted particular industries (Alkali Act 1863) and prescribed substances (Alkali &c. Works Regulation Act 1906). The Alkali Act initially set emission limits on hydrogen chloride from the alkali industry and evolved to apply controls to a wide range of industries and pollutants. As the identity of the activity and the pollutant to be controlled was prescribed in the acts, changes required new acts. Under the Alkali, &c, Works Regulation Act 1906, operators were required to use the best practical means to prevent the escape of noxious or offensive gases.

For the next 80 years, separate environmental regulations developed for the control of pollution to the air, water and land.

A key change in environmental regulations across the European community came with the implementation of the Framework Directive 84/360/EEC on the combating of air pollution from industrial plants. With inputs from UK-based regulators and civil servants, it also transposed into European legislation many of the requirements of UK industrial environmental regulation:

1. Industrial plants listed in an annex had to have prior authorisation from the national or regional authority or regulator before beginning operation or before making any substantial change to the plant.
2. A permit or authorisation could only be issued when the authority was satisfied that all appropriate measures against air pollution have been taken, including:
    a. the application of best available technology not entailing excessive cost (BATNEEC)
    b. the use of the plant that would not cause significant air pollution, particularly of the substances listed in the annex of the directive
    c. when none of the applicable emission limit values were exceeded
    d. when all relevant air quality limit values were taken into account
3. Applications for authorisations and the decisions of the authorities had to be made available to the public.

The implementation of the directive through the Environmental Protection Act 1990 was used to deliver an integrated approach to the regulation of releases to air, land and water for England, Wales and Scotland. The prescribed substances and processes were identified by the Integrated Pollution Control (IPC) Regulations made under the Act. Emission standards for industry sectors were laid out in guidance on BATNEEC drafted in collaboration with the respective industry trade associations.

An integrated approach to the control of pollution to the air, land and water was delivered across Europe by the integrated pollution prevention and control (IPPC) directive (European Council 1996) and was replaced in 2008 (European Parliament and Council 2008). It in turn was replaced by the IED (European Parliament and Council 2010). The IED also incorporated the requirements of several other directives including on emissions from incinerators, combustion plants and titanium dioxide manufacturing. It was implemented in the UK through regulations in 2013 and delivery was achieved by 1 January 2016 (UK Parliament 2013).

The responsibility for delivery of environmental regulation in the UK is divided between the national environment agencies and local authorities. The division is based on past responsibilities/arrangements and the strengths of the regulatory bodies. The position in England is given in Boxes 1.1 and 1.2.

The IED provides a comprehensive framework and level playing field for the environmental regulation of industry across the EU. Its key features are:

1. An integrated approach encompassing the whole environmental performance of the plant including releases to air, water and land, generation of waste, use of raw materials, energy efficiency, noise, prevention of accidents and restoration of the site upon closure.
2. A requirement for a permit to include emission limit values based on the BAT. To define BAT and the BAT-associated environmental performance across the EU experts from member states, industry and environmental organisations, exchange information under the auspices of the European IPPC Bureau of the Institute for Prospective Technology Studies at the EU Joint Research Centre in Seville, Spain. The conclusions as to what BAT is in the resulting BREFs are adopted

Box 1.1 Environment Agency regulated activities (Environment Agency 2015)

Specialism				
	High		Power stations Refineries and fuel processes	Nuclear installations Chemicals Hazardous waste landfill
	↑	Inert waste landfill Inert waste transfer Inert waste treatment Low-impact installations	Metals Cement and other minerals Pulp and paper Tanneries Food and drink manufacture Non-hazardous waste landfill Biowaste Land spreading Agriculture (intensive)	Waste incineration Hazardous waste treatment
	Low	Exempt activities Carriers and brokers registration Hazardous waste producer notification	Metal recovery Non-hazardous waste treatment Non-hazardous waste transfer Agriculture (non-intensive)	
		Low	Risk ⟶	High

by the Commission and become the starting point for setting permit conditions. The IED also sets EU-wide emission limit values for selected pollutants from specified processes including large combustion plants and incinerators.

3. Less strict emission limit values are allowed where an assessment shows that achieving the emission levels associated with BAT would lead to disproportionately higher costs compared to the environmental benefits due to the location, local conditions or technical characteristics of the installation.

Box 1.2 Local authority regulated activities (Department of Environment Food and Rural Affairs 2012)

Refining gas
Metal works, for example, producing pig iron or steel, casting ferrous metal, operating forge hammers or applying fused metal coatings
Melting non-ferrous metals
Surface treating metals and plastic materials
Grinding cement clinker or metallurgical slag
Glass manufacturing
Cellulose fibre reinforced calcium silicate board manufacturing
Ceramic product manufacturing, including roof tiles and bricks
Non-hazardous or animal waste incineration
Manufacturing wood-based boards, for example, plywood
New tyre manufacturing
Disposing of or recycling animal carcasses or waste

4. Member states are required to set up a system of environmental inspections and set inspection plans. There must be site visits at least every one to three years depending on risk.
5. The public has a right to access permit applications, permits and the results of the monitoring of releases, so as to be able to participate in the decision-making process. Emission data must be reported by member states and made accessible to the public through the European pollutant release and transfer register (E-PRTR).

National Regulators

The responsibility for the delivery and enforcement of environmental regulation has shifted over the past 150 years. The style of the regulations, the design of the regulatory bodies and the policy and influence of government have all impacted on the eventual outcome for the environment and process operators.

The original Alkali Acts were enforced by alkali inspectors and a chief alkali inspector answerable to the government. The Alkali Inspectorate survived in various guises and with accountability to different government departments until 1974 when it was subsumed into the newly created

Health and Safety Executive as the Industrial Air Pollution Inspectorate. While there were professional synergies with the staff in the Executive, especially the specialist inspectors, environmental regulation was perceived to be discretionary, collaborative and secretive (O'Riordan and Wynne 1987).

The Royal Commission on Environmental Pollution (RCEP) in its fifth report recommended the creation of a unified inspectorate to deliver an integrated approach to industrial environmental problems (Royal Commission on Environmental Pollution 1976). This call was repeated in its tenth report (Royal Commission on Environmental Pollution 1984). Eventually, responsibility for industrial environmental regulation was consolidated into Her Majesty's Inspectorate of Pollution in England and Wales and Her Majesty's Industrial Pollution Inspectorate in Scotland on 1 April 1987. The Commission had proposed the creation of Her Majesty's Pollution Inspectorate (HMIP) but civil servants baulked at the idea that it was Her Majesty's Pollution. This integrated the Radiochemical Inspectorate and the Hazardous Waste Inspectorate (HWI) with the Industrial Air Pollution Inspectorate, and this was variously seen as a centralising, consolidating or cost-saving action (Andrew Gouldson 2013).

The independence and expertise of the RCEP enabled it to prepare 29 major reports in its 40-year life, till 2011, and significantly influenced environmental regulation. It paved the way for an integrated approach culminating in the 12th report the Best Practicable Environmental Option (Royal Commission on Environmental Pollution 1988). It also made significant contributions on the handling and disposal of waste in its 11th report on Managing Waste: The Duty of Care (Royal Commission on Environmental Pollution 1985). Its recommendation that all those who produce waste should have a "duty of care" to ensure that their wastes were subsequently managed and disposed of without harm to the environment was reflected in the subsequent legislation. The 17th report on the Incineration of Waste called for a national waste management strategy prioritising prevention and reduction of waste, recycling and recovering energy from residual waste (Royal Commission on Environmental Pollution 1993). It reflected a more integrated approach to waste management, which fitted well with the development of integrated regulation and the development of the circular economy.

However, there were cultural difficulties in consolidating the organisations responsible for the implementation of integrated regulation. There was tension between a prescriptive approach to regulation previously seen in waste and water regulation and a more consensual approach seen in industry regulation. Mixed messages and varying levels of engagement with industry resulted in poor-quality permit applications and delays in regulation. Working relationships with industry improved as the 23 guidance notes on the use of the "best practicable means" were replaced by technical guidance drafted in consultation with the respective industries and regulators for processes controlled by both the national and local regulators. This consensual and collaborative approach was later replicated in the development of the European Community BREFs.

Local Authority Regulation

Regulations have been driven by the need to protect health, welfare, ecosystems as well as in response to threats and incidents. The great London smog of 1952, associated with 4703 deaths compared with 1852 deaths in the same period the previous year, led to the 1956 Clean Air Act (Brimblecombe 2002). It prohibited dark smoke from coal burning from chimneys, railway engines and ships and required new furnaces to be smokeless as far as was practicable. Local authorities were empowered to introduce smoke control areas where only smokeless fuel could be burnt.

Noise and smoke from private dwellings, boilers, some industrial plants, steam locomotives and waste burning continues to be regulated by local environmental health officers. Port health authorities have similar powers to local authorities except for control over noise. Local authorities have continued to regulate activities where local control is appropriate and proportionate to environmental impact. This is particularly the case for activities where odour emissions are a potential problem. These include much of the food and drink industry, intensive livestock as well as small combustion plants.

A significant environmental responsibility of local authorities is in local air quality management and the attainment of air quality standards. Where concentrations of pollutants exceed the limit value, local authorities are required to prepare and implement improvement plans. Poor air quality attributable to industrial and other fixed sources can be readily tackled through integrated pollution regulation by the relevant local or national regulator, but the growth in road traffic has been an increasingly problematic source of air quality limit exceedances. European standards have been applied to the design emissions of new vehicles but recent reports indicate that real-life performance can be very different (House of Commons Transport Committee 2016).

Where local authorities have declared traffic-related air quality management areas they could undertake or require emissions testing on vehicles driven in, through or out of it. The level of action and enforcement of these environmental controls has been significantly impacted by the severe reduction in funding of local authorities and the consequential reprioritising of their work.

Greater London has already implemented and other cities are considering limited low-emission zones restricting the vehicles that can drive through designated parts of cities. Significant improvements to local air quality are only likely to be achieved when there is more coordinated action between local authorities, central government and national regulators.

Waste and Water Management Regulation

Waste management regulation emerged out of the duty of public authorities aimed at protecting public health. Powers to remove household and commercial waste were given to local authorities in the 1936 Public Health Act. The Town and Country Planning Act 1947 required waste disposal sites to have planning permission, the precursor for the permitting of waste activities. Until the 1960s, the national approach to waste management was bury or burn with landfill sites designed to dilute and disperse the liquids which leached from them. This was similar to the dilute and disperse tall chimney solution for many emissions from industry especially power stations.

The discovery of 36 drums of sodium cyanide ash in disused clay workings near Bermuda village, Nuneaton, in February 1972 and other reports of the dumping of hazardous waste highlighted the failure of the existing voluntary code of practice and led to the Deposit of Poisonous Waste Act 1972 (Hansard 1972). The Act introduced a general prohibition on depositing poisonous and other dangerous wastes, civil liability and a duty to notify responsible authorities before removing or depositing waste. It also introduced penalties of imprisonment for up to five years and unlimited fines. The 1974 Control of Pollution Act introduced the licensing of waste activities to prevent water pollution, protect public health and prevent local detriment. Part II of the Environmental Protection Act 1990 (England, Wales and Scotland) (EPA 1990) and similar legislation in Northern Ireland tightened up the licensing requirements. The act as amended:

1. prohibited the unauthorised or harmful depositing, treatment or disposal of waste
2. provided for a Duty of Care as respects waste
3. required Waste Management Licences

In August 1983, a Hazardous Waste Inspectorate (HWI) was established for England and Wales to oversee the disposal of hazardous wastes and the operation of waste disposal sites. Following concerns that local authorities were both operators and regulators of waste facilities, their regulatory and disposal functions were separated by the EPA 1990. Waste is now managed through national permitting schemes and regulation directed by a series of European Waste Framework Directives.

The first Public Health Act in 1848 provided for the local authority management of the provision of water and the control of sewage. Local bodies were responsible for the protection and management of the rivers that fed towns and cities as well as supporting industrial activities. The local river authorities continued until the Water Act 1973, which consolidated the authorities in England and Wales into regional water authorities. Their responsibilities included water conservation, supply, sewage and its treatment, water pollution control, drainage, fishing and flood prevention. Under the Water Act 1989, the water supply and waste water management functions were separated off into commercial water companies. The National Rivers Authority took over the remaining responsibilities until the creation of the Environment Agency (England and Wales) in 1996, with separate regulation of drinking water by the Drinking Water Inspectorate.

In Scotland, the River Purification Authorities continued to regulate water pollution until the creation of the Scottish Environment Protection Agency (SEPA) on 1 April 1996. The Water Industry (Scotland) Act 2002 established the publicly owned Scottish Water to provide water management services and the Drinking Water Quality Regulator for Scotland.

The Water and Sewage Services (Northern Ireland) Order 2006 (SI 2006/3336, NI 21) established water regulation arrangements in Northern Ireland similar to England, Wales and Scotland. A separate Drinking Water Inspectorate is responsible for regulating drinking water quality. Northern Ireland Water provides water and waste water management and is owned by the Northern Ireland government.

In England, Wales, Scotland and Northern Ireland, water regulation is increasingly being combined into an integrated pollution and prevention approach managing releases to air, water and land.

The UK regulators coordinate their work and share knowledge through the Belfast Group which includes representatives of the regulators and sponsoring government departments. The UK regulators have also been active members of IMPEL, the EU Network for the Implementation and Enforcement of Environmental Law, which is made up of organisations or authorities working in the public sector who implement and enforce environmental legislation.

Working Arrangements

The collaboration between regulators and many regulated industries has improved significantly over the past 30 years as environmental regulation has become more distinct from health and safety or local nuisance regulation. In 1989, the then Chief Inspector of HMIP encouraged an "arm's length" relationship with industry signalling the end of an alleged cosy relationship with industry and possibly reflecting a lack of trust and concerns about the performance of particularly the radiochemical industry.

A more general suspicion of industry by the public and politicians is also reflected. However, the need for more comprehensive permit applications under the new IPC regulations resulted in pre-application discussions and agreements on the best technologies (Ends 1992), and the "arm's length" approach was dropped in 1993.

IPC saw the involvement of consultants to help with permit applications, undertaking dispersion modelling, release analysis and environmental impact assessments. The Environmental Auditors Registration Association was formed in 1992 as a result of the development of environmental professionals and their need for recognition as a distinct profession. It merged with the Institute of Environmental Management in 1999 to form the Institute of Environmental Management and Assessment (IEMA). IEMA has grown to become the largest sustainability professional membership body in the UK, promoting best practice standards in environmental management and auditing for industry, public, private and non-governmental sectors.

The need for and development of professional competency standards in the waste management industry were led by the Chartered Institution of Wastes Management (CIWM) established in 1898. It seeks to raise standards for those working in and with this very diverse sector that has been undermined by less scrupulous operators. CIWM produces best practice guidance, develops educational and training initiatives and provides information on key waste-related issues.

In 1987, the Institution of Public Health Engineers (founded 1895), Institution of Water Engineers and Scientists and the Institute of Water Pollution Control formed what is now the Chartered Institution of Water and Environmental Management. Its focus is on the development of the science and practice of water and environmental management for the public benefit.

During the 1990s, the professional organisations above, and others, identified the need for a strong independent body to champion and provide recognition for environmental professionals. The Society for the Environment (SocEnv) was started in 2002 by eight professional bodies and by the time its Royal Charter was granted in 2004, this had risen to ten. SocEnv comprises 24 licensed professional bodies and learned societies, representing between them over 400,000 practitioners working in a wide range of professions, sharing a vision—sustainability through environmental professionalism. The constituent bodies are licensed by SocEnv to award the Chartered Environmentalist qualification to their members.

Working together, the professional bodies are able to provide a strong and independent voice for environmental protection and sustainable development. This has been missing since the impartiality and strength of the environmental regulators has declined as they came under closer control by their government sponsors.

Development of Integrated Regulation and the Formation of the Environment Agencies

In the late 1980s and early 1990s, there was a desire in government, civil service and industry to see more unified regulation as exemplified in the USA by their Environmental Protection Agency (EPA) formed in 1970. Action was also needed to address increased pollution incidents and the difficulties HMIP were experiencing in delivering IPC through the EPA 1990. One of the issues was the need to engage with a number of statutory consultees, which was resource intensive and who either made significant demands or after delay raised no requirements at all. Integration and better coordination were required. In response, the Environment Agency (England and Wales) and the SEPA, created by the Environment Act 1995, came into existence on 1 April 1996 as separate executive agencies of government. They took over the pollution, water and waste regulation roles and responsibilities from several predecessor organisations.

In the early years of the agencies, strenuous efforts were made to reconcile the more detached standard approach of water regulation with the policing approach in waste regulation and the collaborative approach of industry regulation. In parallel, there was a strong demand from government, industry and public bodies for the public demonstration of improved performance (Ends 1996). For regulated industry, this was delivered through the development of the Pollutant Release and Transfer Register (PRTR) better known as the Pollution Inventory, identifying the emitters/sources of the greatest amount of a wide range of polluting substances in air, land and water, industry compliance and improvements in performance (Ends 1999). The register has been revised several times and is updated annually. In 2017, it provided data on the emitters from industry and agriculture of 91 polluting substances as well as waste transfers. In March 2017, the government announced a review of the register but it was unclear whether this was just to meet UN reporting requirements or a precursor to a reduction in the scope and application of the register as part of the move to reduce the reporting requirements on business (Department for Environment, Food and Rural Affairs 2017). The UK is committed to the scheme as a signatory of the Kiev Protocol to the Aarhus Convention.

In 2008, the Northern Ireland Environment Agency was formed as a conservation agency within the Department of Agriculture, Environment and Rural Affairs. Since then, there have been calls for the creation of an independent EPA (Department for the Environment Northern Ireland 2015). On 1 April 2013, part of the Environment Agency covering Wales was merged into Natural Resources Wales leaving the Environment Agency responsible for regulation only in England.

The objectives and priorities of the UK environment regulators are shown in Box 1.3.

Box 1.3 UK Environmental regulators' objectives and priorities

Natural Resources Wales

To pursue the sustainable management of natural resources in relation to Wales and apply the principles of sustainable management of natural resources:

1. Embedding Sustainable Management of Natural Resources
2. Delivering Customer Focus
3. Improving Efficiency and Service Delivery
4. Developing our People and Teams
5. Developing Enterprise and Business
   (Natural Resources Wales 2016)

Northern Ireland Environment Agency

To create prosperity and well-being through effective environment and heritage management and regulation

1. Delivering effective compliance and implementation of legislation and international obligations

2. Improving understanding and appreciation of our environment
3. Supporting a sustainable economy
4. Delivering reformed and effective planning
(Department of Agriculture, Environment and Rural Affairs 2016)

Environment Agency

To create better places for people and wildlife, and support sustainable development

1. Working with businesses and other organisations to manage the use of resources
2. Increasing the resilience of people, property and businesses to the risks of flooding and coastal erosion
3. Protecting and improving water, land and biodiversity
4. Improving the way we work as a regulator to protect people and the environment and support sustainable growth
(Environment Agency 2016)

Scottish Environment Protection Agency

To protect and improve the environment in ways that, as far as possible, also help create health and well-being benefits and sustainable economic growth.

1. Regulating activities to control their impact on the environment and human health
2. Delivering environmental improvements
3. Safeguarding communities
4. Identifying and responding to new environmental challenges
5. Understanding the state of the environment and its impact on human health
6. Promoting positive environmental behaviours
7. Acting to combat climate change
8. Championing sustainable resource use
9. Enabling delivery of high quality, consistent and customer-focused services
10. Developing and retaining high performance people
(SEPA 2016)

1.4 Risk-based Regulation

In August 1997, the Environment Agency introduced an Operator and Pollution Risk Appraisal (OPRA) scheme to help it allocate inspection resources more effectively on a site-specific basis. Its earlier development by HMIP was driven by the need to improve transparency in the application of charges, which had previously been based on standard inspection frequencies and risks for each regulated sector. The scheme assessed the operator's performance and the intrinsic pollution potential (Environment Agency 1997). In 2000, a similar scheme was developed for waste activities.

In 2001, the Environment Agency introduced Operator Monitoring Assessment (OMA) to assess the quality of companies' self-monitoring and to enable it to target its own monitoring on sites with poorer scores. It was similar to the OPRA scheme and reduced the overall amount of independent monitoring with the best operators receiving no monitoring visits from the Agency. It was welcomed by industry as most emission monitoring was done by companies with the Agency auditing their data through independent check monitoring by consultants and charged to the operators (Ends 2001).

In 2002, the Environment Agency revised the OPRA methodology to use it to set fees for pollution prevention and control (PPC) and waste licensing. Importantly, the system gave credit to companies with certified environmental management systems (Environment Agency 2002). A similar scheme was developed by SEPA.

Application and subsistence changes continue to be based on the regulatory effort required which in turn is determined by the operator performance, scale/nature of the activity and environmental impact. (Full details of the current OPRA scheme for England can be found on the Environment Agency's website.)

The UK regulators have extensive responsibilities, objectives and priorities, which have evolved in anticipation of and in response to a combination of environmental issues, political pressure and public perceptions. In particular, contributing towards sustainable development and specifically economic development has become the core objective. They have built on the different cultures of their predecessor organisations while at the same time taken a more holistic approach to environmental regulation. Other organisations have also looked for new ways of delivering regulation.

The National Society for Clean Air and Environmental Protection (NSCA) Commission on Industrial Regulation and Sustainable Development reported on Smarter Regulation in 2001, linking effective pollution control with sustainable development (NSCA Commission on Industrial Regulation and Sustainable Development 2001). It considered alternatives to regulation: voluntary initiatives, economic instruments, negotiated agreements and adapting regulation to meet individual business needs. It concluded that:

1. Regulation alone cannot deliver sustainable development.
2. Traditional site-specific regulatory systems, in particular, are unlikely to prove adequate to the needs of sustainability.
3. Environmental protection systems need to be reformed to make them more objective led and more focused on actual environmental outcomes.
4. There was a need to look beyond current systems by:

    a. moving the focus of regulation from processes towards integrated regulation of products
    b. developing policies to promote environmental innovation and resource productivity
    c. recognising the impact of new economy information and communications technologies

In 2003, the Environment Agency proposed to give chemical companies with low environmental impacts a lighter regulatory touch under the IPPC (Environment Agency 2003). However, there was little interest from operators, as the application process was not significantly less onerous than that required for a full application.

A significant change in the relationship between regulators and business occurred in 2003 with the publication of the REMAS study by the Environment Agency (England and Wales) on the benefits of the use of environmental management system standards. Operators of regulated processes were required to have a written management system although the international standard ISO 14001 did not specify that certified businesses must be compliant with regulations. The REMAS study showed that management standards drive environmental improvement but do not necessarily improve compliance (Environment Agency R&D Technical Report P6-017/2/TR 2003; www.remas.info 2006). It was therefore in the interests of a site regulator to look at the management system and any audit against it as part of the inspection. Non-compliances with the system were a good starting point for investigation.

In 2004, the Environment Agency published Delivering for the Environment: A 21st Century Approach to Regulation, in which it developed its proposals for modern regulation (Environment Agency 2004):

1. defining outcomes and risks
2. choice of instruments
3. compliance assessment and enforcement
4. evaluation and the provision of information

In addition to regulation through bespoke and standard permits, registrations and the direct application of legislation, it identified alternative approaches:

1. environmental taxes
2. trading schemes
3. negotiated or voluntary agreements
4. education and advice
5. environmental management schemes

In 2006, Compliance Assessment Plans (CAPs) were rolled out by the Environment Agency as part of an auditing programme and specified the amount of regulatory activity a site should receive to ensure compliance (Ends 2006). The CAPs were aligned with the scores from the OPRA scheme.

The UK environmental regulators maintained a high level of dialogue with the national trade bodies for the regulated industries. HMIP and its predecessors even had formal documented minutes of meetings of these industry liaison groups and its internal groups. As more industries were brought under IPC and IPPC the number of sector groups increased. Staff from all the regulators with direct experience of the regulation of the various sectors worked together in the sector groups led by a regional regulator or national policy manager. Sector Plans were developed in collaboration with the trade associations for each separate industry sector. They identified the most significant impacts of the sector and set long-term improvement objectives and performance indicators. The aim was to target poor performing sites with in-depth audits which also resulted in a drop in the overall level of inspection visits. There was concern about documenting the meetings with industry often resulting in separate records being held by industry and regulators as well as a brief common text.

In 2006, the House of Commons Environment, Food and Rural Affairs committee reviewed the effectiveness, funding and relationships of the Environment Agency (House of Commons Environment, Food and Rural Affairs Committee 2006). It recognised the success of its performance as a "modern regulator" balancing national consistency with local flexibility but challenged its consistency between the policy centre and inspectors on the ground. It reflected the concerns of stakeholders that it struggled to combine its regulatory role with that of the "Champion of the Environment".

In November 2016, following the record rainfall and storms in winter 2015–2016 which disrupted communities across northern parts of the UK, the committee recommended a new model for managing food risk (House of Commons Environment, Food and Rural Affairs Committee 2016). Their view was that the current flood risk management structures were fragmented, inefficient and ineffective. The committee recommended establishing a new National Floods Commissioner for England. Delivery would be via a new English Rivers and Coastal Authority, taking over the current Environment Agency roles on national flood risk management. New Regional Flood and Coastal Boards would coordinate regional delivery of national plans, in partnership with local stakeholders, taking on local authority roles.

This splitting of the work of the Agency could significantly reduce its ability to shift resources where required in response to national and local incidents.

Environmental Management Systems and Compliance

In 2010 the Environment Agency trialled its voluntary Environmental Permitting Regulations Assurance Scheme (EPRAS) whereby operators submitted an Annual Compliance Statement (ACS) signed at a senior level. The assurance provided by high-level ownership of environmental regulatory matters was designed to give the regulator the confidence to reduce the level of direct regulatory inspection and audits of that company. The Mineral Products Association representing the cement industry was a keen advocate. The risks associated with the approach were seen by the industry as:

1. increased liability for senior executive managers
2. increased scrutiny arising from publication of statements
3. increased sanction for failure to fulfil requirements
4. the need to put in place additional internal systems in order that senior managers had sufficient confidence to sign their statements

Combining the use of management systems and the ACS to provide an integrated approach was intended to enable the operator to earn autonomy from regulation. Increasing performance and a higher level of assurance with permit compliance would facilitate less direct regulatory intervention from the regulator who had greater confidence in the information supplied. Combining a third-party audited management system, an operator with good performance, and one who produces an ACS, should give the greatest level of confidence of compliance to the regulator. Where operators do not use accredited systems (e.g. small and medium size enterprises (SMEs), where use of an accredited environmental management system (EMS) may be too onerous), other checks could be required. Regulatory control would revert to normal if there was poor performance, prosecution or formal caution, false or misleading information or if information was not provided on request.

The approach should provide better evidence of compliance and enable the regulator to better target and reduce the number of audits and inspection, which would be determined on a sector basis. Unannounced inspections and audits would still feature as well as investigations of incidents or complaints.

A report by Ends in 2012 found that organisations with a strong internal commitment to improving performance could gain real value from certification to green standards (Ends 2012). While companies were keen on the badge, there was often far less commitment to ensuring that the certified environmental management system actually delivered improved performance and good compliance. It was also apparent that the certification bodies were not holding organisations to account well enough to ensure good results. This may be due in part to the recognition that it was the company that selected and paid the certifier.

Research by the Scottish and Northern Ireland Environment Agencies has identified difficulties between environment agency inspectors and management standard auditors. The certification bodies did not consider their role to be ensuring compliance (Planet and Prosperity Ltd 2013).

A review in 2014 by the Environment Agency found that during the EPRAS trial there was no deterioration in overall average site compliance performance compared to the preceding two years and no increase in substantiated complaints relating to the sites during the trial (Environment Agency 2014a). While participants felt that the introduction of an assurance scheme would reduce the costs of regulation through reduced inspections, the scheme did not significantly reduce administrative burdens or costs. However, some operators valued the face-to-face contact with Environment Agency inspectors and wanted to maintain the contact. Operators said a lower subsistence fee was not a big driver for taking part. They felt that other Environment Agency and wider government schemes were often a greater burden than those from complying with their EPR permit. As a result, the Environment Agency invited regulated industries in England to take part in a wider earned recognition scheme that could give them positive publicity for complying with environmental rules.

The Royal Society for the Protection of Birds in its report on the use of voluntary approaches such as industry self-regulation as an alternative to mandatory rules and regulations found that the impacts of most voluntary schemes were limited. The efforts of responsible businesses were often undermined by the failure of such schemes to attract widespread industry participation and compliance (McCarthy and Morling 2015).

The SEPA in its annual operating plan for 2016/2017 introduced a business sector and a more consensual approach to regulation. While there were more concrete objectives for the water environment, numeric targets to improve operator compliance were removed. There were commitments to reduce the number of non-compliant sites and SEPA also expected regulated businesses to conduct more self-monitoring.

Looking to the future, the ambition must be that data from self-monitoring is available online in real time. This should boost the public's confidence in environmental regulation as there is a general concern that results from self-monitoring will be less reliable. There have been three prosecutions because of falsified self-monitoring results: St Regis Paper in 2010 (Ends 2010b), Sevalco in 2004 (Ends 2004) and Leigh Environmental in 2000 (Ends 2000).

Deregulation

Deregulation, better regulation or less regulation have been a policy mantra for successive UK governments and the EU over the past few years. However, the initiative has not always been seen to benefit the environment (Christine Berry 2015). In April 2011, the government published its One-in, One-out: Statement of New Regulation as part of its commitment to reducing the cost and volume of regulation on the economy (Department for Business, Innovation and Skills 2011a). This reiterated the requirement from autumn 2010 on each government department to assess the net cost to business of complying with any proposed regulation, secure validation of the net cost to business and find a deregulatory measure that relieved business of the same net cost. It introduced a requirement for sunset clauses to be included in new regulations and stopped over-interpretation (gold plating) of the EU law. It also initiated a wide-ranging review of existing regulations.

This reflected concerns about regulatory burden by many politicians, interest groups and in the media (Federation of Small Businesses 2011; Department for Business, Innovation and Skills 2011b). Less was said in the media about the benefits of regulation, the need for regulation or drive for new and better regulations. Instead, existing environmental regulations have been challenged (Department for the Environment, Food and Rural Affairs 2014).

Resistance and challenges to environmental regulations and regulatory change have taken different forms. In September 2011, the UK government announced that the review of environmental regulations will be the focus of the "Red Tape Challenge" (Department for Environment 2011), causing significant misgivings amongst trade and green groups. However, the review by DEFRA identified relatively few laws that were obsolete (Department for Environment 2012). The regulators were under pressure to identify regulatory and administrative requirements on businesses that could be reduced. This caused significant distraction from the primary task of environmental protection.

The Environment Agency published its response to the pressure for deregulation with an update to its approach to modern regulation, Delivering for the environment in 2014 (Environment Agency 2014b). The paper explained why regulation was important and the benefits it had helped to achieve. It also justified the Agency's regulatory role, principles to which the Agency worked, the approaches and tools used, working with others and aims for future work. Against a background of funding and staff cuts alongside its fellow UK regulators, it has to work hard to justify and explain its work.

When the outcome of the red tape challenge was announced in January 2015 there was strong emphasis on helping businesses meet their environmental obligations—"by March 2015 Defra will have slashed 80,000 pages of environmental guidance saving businesses around £100 million per year" (GOV.UK 2014). What was not stated was that much of the guidance had been drafted in collaboration with business to help delivery of regulations or the basis of the claimed savings.

In response, the manufacturers' organisation, the EEF, stated that what they wanted was simpler data reporting and better regulations and guidance but without compromising the levels of environmental protection afforded by it (EEF 2015). According to the World Economic Forum of the G7 group of countries, the UK is the least regulated country even though there have been increased efforts and pace since the new UK government in 2010 (Schwab 2015).

Transfer of existing guidance to the central government website or to national archives has not helped improve its accessibility. In its acknowledgement of the concerns of the waste and recycling industry to proposed deregulation, the government pledged to improve enforcement and guidance (Department for Business, Innovation and Skills 2016).

On 3 March 2016, the Business Secretary Sajid Javid announced at the British Chamber of Commerce Conference new plans to cut red tape of "one in, three out". "If departments want to bring in new regulatory costs for things that weren't in our manifesto, they will be expected to find savings worth three times as much." While attractive on paper it became an even bigger disincentive than the earlier one in, one out to civil servants and politicians to improve or replace existing regulations, because of the work required.

1.5 Policy Perspectives on the EU and the UK Approach to Regulation

Much of UK and EU environmental regulation has been based on the principle of command and control, with what is to be controlled in terms of activity or pollutant and how it is to be controlled in terms of control mechanisms and what the outcome must be, in terms of levels of pollutant release, impact on the activity or on the receiving environment. Over the past 30 years, three broad approaches to environmental regulation have evolved (François Lévêque 1998):

1. regulatory instruments, whereby public authorities mandate the environmental performance to be achieved, or the technologies to be used, by firms
2. economic instruments, whereby firms or consumers are given financial incentives to reduce environmental damage
3. voluntary approaches, whereby firms make commitments to improve their environmental performance beyond what the law strictly demands

Analysis of UK and EU policies shows that voluntary approaches are in the ascendancy. A good example of a voluntary approach is the Chemical Industries' Responsible Care Programme (International Council of Chemical Associations 2016). It is a global initiative designed to secure continuous improvement and achieve excellence in environmental, health, safety and security performance. It is also part of the industry's commitment to sustainable development. It was launched in 1985 by the Canadian Chemistry Industry Association of Canada and is now practised by the industry globally. Company CEOs are expected to take responsibility for its delivery.

Voluntary approaches, where a business or industry has of its own volition made a significant change to its activities or environmental impacts without any external pressure or influence, unless there have been substantial cost savings, are harder to find. The exception to this is the number of businesses that are reducing their emissions of greenhouse gasses as a result of national and international agreements on addressing the causes and impacts of climate change. This is often accompanied by significant communications to the market, customers, public and staff generating a good public relations impact on the business.

Negotiated agreements are increasingly common across Europe. In 2001, the UK government signed climate change agreements with 48 industry and commercial sector organisations. The agreements allowed the signatories an 80% discount on the climate change levy in return for legally binding commitments to reduce emissions and/or improve energy efficiency. In 1999, the European Commission recommended a voluntary agreement with some global car manufacturers on the energy efficiency of private vehicles (The Commission of the European Communities 1999). Most voluntary agreements, once concluded, are backed up by sanctions for non-compliance (Silvia Rezessy et al. 2005).

When DEFRA undertook a comprehensive audit of its environmental regulations, it found that 53% were of domestic origin and 47% of EU origin (Department for Environment, Food and Rural Affairs 2015). DEFRA concluded that the general benefits of its environmental regulations outweighed the net cost to business and civil society by at least three to one.

Whereas UK legislation and policies can and do change within months, depending on the views and position of ministers and the government, EU legislation generally requires the agreement of 28 countries as well as facing scrutiny by the European Parliament. As a result, it can take significantly longer to develop and implement, but once implemented can withstand sustained challenges by interest groups and national governments. Since the UK joined the European Communities on 1 January 1973 the influence of and UK impact on the development of European environmental legislation can be clearly seen in the wording of directives and regulations.

This is due in significant part to the knowledge and background of the UK experts who have worked in Brussels on the legislation.

The first United Nations Conference on the Environment in Stockholm in 1972 initiated the United Nations Action Programme and, amidst growing public concern about limits to growth including the report by the Club of Rome, (Donella H. Meadows et al. 1972) established the foundations for what is now encompassed by "sustainable development".

Article 191 of the Lisbon Treaty (European Commission 2012) sets out the objectives of the EU's environment policy:

1. "Union policy on the environment shall contribute to pursuit of the following objectives:

   a. preserving, protecting and improving the quality of the environment,
   b. protecting human health,
   c. prudent and rational utilisation of natural resources,
   d. promoting measures at international level to deal with regional or worldwide environmental problems, and in particular combating climate change."

2. "Union policy on the environment shall aim at a high level of protection taking into account the diversity of situations in the various regions of the Union. It shall be based on the precautionary principle and on the principles that preventive action should be taken, that environmental damage should as a priority be rectified at source and that the polluter should pay."

"In this context, harmonisation measures answering environmental protection requirements shall include, where appropriate, a safeguard clause allowing Member States to take provisional measures, for non-economic environmental reasons, subject to a procedure of inspection by the Union."

The European Communities adopted their first European Environmental Action Programme in 1972 based on the ideas that prevention is better than cure and the "polluter pays" principle. The aim of the community environment policy was to "improve the setting and quality of life, and the surroundings and living conditions of the peoples of the Community" (European Community OJ C112/1 1973). The objectives of the programme were to:

1. prevent, reduce and as far as possible eliminate pollution and nuisances
2. maintain a satisfactory ecological balance and ensure the protection of the biosphere
3. ensure the sound management of and avoid any exploitation of resources or of nature which causes significant damage to the ecological balance
4. guide development in accordance with quality requirements, especially by improving working conditions and the settings of life
5. ensure that more account is taken of environmental aspects in town planning and land use
6. seek common solutions to environment problems with states outside the community, particularly in international organisations

The third (1982) and fourth (1987) action programmes strongly linked the European internal market and environmental policies including the positive economic benefits gained from environmental policies (Hey 2005). Emission standards and controls, including emission limit values, needed to be harmonised to avoid distorting competitiveness. The UK supported the adoption of environmental quality standards. Waste avoidance and an integrated approach to pollution control were developed. The use of indirect, economic instruments was promoted by the European Commission in 1989 displacing the previous command-and-control approach and resulted in the carbon dioxide/energy tax that piloted in May 1992.

The fifth action programme (1992) confirmed the principle of sustainable development (Council and the Representatives of the Governments of the Member States 1993). It took a sector-based approach targeting the most polluting industries such as large combustion plants and incinerators and setting long-term pollution reduction objectives. It supported the use of financial incentives and voluntary instruments. More ambitious proposals on areas including energy and packaging were watered down. A new regulatory agenda emerged driven by framework directives, voluntary agreements and self-regulatory management tools. It marked a shift from Europe-wide standards to consensus and local agreements with industries.

The sixth action plan (1997) developed the collaborative approach with industries, including integrated product policies, the wider use of environmental standardisation, voluntary agreements and collaboration between experts (Commission to the Council, the European Parliament, the Economic and Social Committee and the Committee of the Regions 2001). The European Commission shifted from being an initiator of legislation to managing processes and the development of policy. The action plan introduced the concept of Thematic Strategies, as a modernisation of EU environment policy making, taking a broader, strategic approach. Thematic Strategies developed an integrated approach and were key elements of the Commission's Better Regulation strategy: they were accompanied by an impact assessment covering the economic, social and environmental impacts of the different options. They were the result of extensive stakeholder consultations and aimed at simplifying the existing regulatory framework. Environmental regulation became standardised across the European community for example through the implementation of Council Directive 96/61/EC of 24 September 1996 concerning IPPC

The seventh Environment Action Programme (EAP) entered into force in January 2014 and guides European environment policy until 2020. It includes a vision of where it wants the Union to be by 2050 (European Union 2014):

> In 2050, we live well, within the planet's ecological limits. Our prosperity and healthy environment stem from an innovative, circular economy where nothing is wasted and where natural resources are managed sustainably, and biodiversity is protected, valued and restored in ways that enhance our society's resilience. Our low-carbon growth has long been decoupled from resource use, setting the pace for a safe and sustainable global society.

The action plan identifies three key objectives:

1. protect, conserve and enhance the Union's natural capital
2. turn the Union into a resource-efficient, green and competitive low-carbon economy
3. safeguard the Union's citizens from environment-related pressures and risks to health and well-being

Rather than identifying the legislation as it might have done in the past, it sets out more generic means to achieve the goals:

1. better implementation of legislation
2. better information by improving the knowledge base
3. more and wiser investment for environment and climate policy
4. full integration of environmental requirements and considerations into other policies

These objectives require that markets reflect the real cost of the environment:

1. applying the polluter pays principle more systematically
2. phasing out environmentally harmful subsidies
3. shifting taxation from labour towards pollution

Cutting across these actions it also identifies two priority objectives:

1. to make the Union's cities more sustainable
2. to help the Union address international environmental and climate challenges more effectively

The member countries are committed to:

1. update air quality and noise legislation
2. improve implementation of legislation relating to drinking and bathing water
3. tackle hazardous chemicals, including nanomaterials, that interfere with the endocrine system and chemicals in combination, as part of a broader, strategic approach for a non-toxic environment

The action plan seeks better integration of environmental concerns into other policy areas, such as regional policy, agriculture, fisheries, energy and transport. This will ensure that there is better decision-making and coherent policy approaches that deliver multiple benefits. The action plan should also ensure an expanding market for environmental goods and services.

The action plans have changed the style of directive and regulation. European and other international agreements and regulations have significantly changed the way the environment has been protected and improved.

Past differing local and national standards, limits, regulations and enforcement have become more consistent across Europe through the implementation of comprehensive directives such as the EU Directive 2010/75/EU on industrial emissions (IPPC) published in 2010 for implementation by 2013. As a result, transposition into UK legislation and delivery by businesses and regulators have been significantly smoother than the implementation of earlier, more prescriptive European legislation.

International agreements and conventions have been implemented across Europe through a consistent standard approach. For example, Regulation (EC) No 1013/2006 of 14 June 2006 on shipments of waste lays down procedures for the transboundary shipments of waste. It implements the provisions of the "Basel Convention on the Control of Transboundary Movements of Hazardous Wastes and Their Disposal".

Environmental regulation and practices have changed as a result of problems with particular industries or the development of new industries.

The interface between government policy and environmental protection was demonstrated in the privatisation, and subsequent sales of the electricity-generating power stations in 1990s, which coincided with the implementation of IPC and the environmental permitting of power stations. There was no way that the privatisation was going to be jeopardised by the immediate imposition of tight controls on sulphur dioxide and nitrogen oxide emissions requiring additional expensive pollution controls. Emission concentration and mass limits were established for each plant to protect the local environment, and company-wide mass emission limits agreed to minimise the long-range impacts. Initially, this resulted in internal emissions trading within the principle companies but as plants were sold the site allocations became very valuable. The importance of emissions controls came to a head when National Power tried to sell Drax power station with insufficient pollutant allocations meaning that in theory it could not be operated by its new owners until it received a new allocation from its regulator (Ends 1998).

The development of onshore hydraulic fracturing (fracking) has highlighted the challenges facing the environmental regulator. The UK government position at the beginning of 2016 was that shale gas had the potential to provide the UK with greater energy security, growth and jobs. Through a variety of incentives, it encouraged exploration and development. However, the devolved administrations in Scotland and Wales had a moratorium on unconventional gas and oil so there was pressure to develop in England.

Regulations forbid fracking in national parks, the Norfolk and Suffolk Broads, areas of outstanding natural beauty and world heritage sites, but fracking can be undertaken up to 1.2 km underneath them, via horizontal drilling (Department of Energy and Climate Change 2016a). Fracking is also banned in protected groundwater source areas.

Before any activity can start the potential operator must secure planning permission from the local planning authority. It has to decide whether the activity is acceptable at that particular location, after local communities and other interested people have had the opportunity to comment on the proposal. There is often significant pressure and a tendency to want to include environmental protection measures at this stage. The operator must also apply for a permit to the Environment Agency and the details of the application have to be publicised. The regulator has to take account of and where necessary include conditions in the permits: to protect water resources, including groundwater (aquifers); to approve the chemicals that form part of the hydraulic fracturing fluid; to ensure treatment and disposal of mining waste including any naturally occurring radioactive materials; and to protect air quality.

The regulator cannot challenge the need for the proposal or take account of wider climate change aspects. This often leads to significant local and national frustration that more substantial environmental issues are not being addressed. This frustration repeats itself with other projects which can have a significant local impact on the quality of life including power stations, waste management facilities and mineral extraction.

1.6 An Overview of the Scope and Likely Future Development of Regulatory Policy

EU membership has had a very significant impact on UK environmental regulation policy. In the 1970s, the UK had a reputation as the "Dirty Man of Europe". As a result of the demand for cheap electricity, the political position and power of the Central Electricity Generating Board, the UK had the highest sulphur dioxide emissions in Europe. A policy of dilute and disperse applied to polluting discharges to air and water.

EU membership has had a dramatic impact on the standards and the enforcement of environmental regulation. Future regulatory policy will therefore continue to be strongly influenced by the relationship between the UK and the rest of Europe. It will also be affected by the development of international agreements, although there is often a significant difference between governments making an international commitment and the timing and actions required to implement it. This is well demonstrated in the enthusiasm for the Paris Agreement on climate change and the actions by signatories, particularly the UK to deliver the commitments.

When the coalition government started its work in May 2010 the new Prime Minister David Cameron announced that "I want us to be the greenest government ever—a very simple ambition and one that I'm absolutely committed to achieving" (Ends 2010a).

This announcement heralded a period of significant change in environmental regulation which was not always consistent with the declared ambition. Government environmental initiatives have come and gone and come back again such as the charge on the use of a plastic carrier bag (Statutory Instruments 2015). Energy policy has fluctuated from support for renewable energy to almost complete withdrawal and priority given to fracking and nuclear generation. The deregulation drive has seen the loss of a wide range of environmental standards and guidance. Increasingly, the environment has become a political football played according to perception and prejudice.

This was well demonstrated in 2015 when Amber Rudd, the Energy and Climate Change Secretary, stopped new subsidies to onshore wind farms on the grounds that the technology should stand on its own feet and save bill payers' money. However, the policy position on onshore wind farms had been shifting from February 2012 when 101 backbench Conservative Party MPs wrote to David Cameron demanding "dramatic cuts" in subsidies. While a moratorium was rejected by the coalition partners, the Conservatives went into the 2015 election promising to "halt the spread of onshore wind farms". The reality is that onshore wind continues to attract significant investment and can be the most cost-efficient way of producing low-carbon energy (Frankfurt School—UNEP Collaborating Centre for Climate & Sustainable Energy Finance 2016).

Political intervention in environmental regulation was well demonstrated by the selection process for the new chief executive of the Environment Agency for England. After an initial open process, the final appointment was through and within the government civil service. The challenge to the new appointee was whether he would be able to rise above political expediency and seek the best solution for the environment (Kaminski 2016).

The future development in environmental regulation is more likely to be driven by the devolved administrations and local authorities.

Regional cities and towns have always had a significant interest in local environmental issues experiencing the direct impacts of poor air quality previously attributed to industry and now primarily due to traffic. Existing devolution of environmental responsibilities has helped. The Scottish government has been a strong supporter of renewable energy with significant onshore and offshore wind farm developments making up for the resistance in Westminster. The Welsh government is at the forefront of commitments on sustainable development. In Northern Ireland the government is updating its industrial and waste regulations.

The Mayor of London has a responsibility for major roads and transport as well as being the final arbiter on planning. This has led to significant environmental successes such as the introduction of the congestion charge in central London and the low emission zone covering most of Greater London by Ken Livingston. Boris Johnson in 2010 substantially reduced the size of the congestion zone in the city and delayed the introduction of the low emission zone, a move which exacerbated the serious problem of poor air quality and breach of EU standards. He did invest heavily in improvements for cycling and the bus fleet, backing an ultra-low emission zone for central London. The new Mayor Sadiq Khan is already challenging the government's air quality policies.

Through "City deals", cities receive an annual capital sum to spend as they see fit rather than having to plead for money. The first deals were published in 2012 and many contained environmental themes and objectives. Directly elected mayors have significant powers on health, planning, transport and environment. New combined authorities will have responsibility for public transport, skills, regeneration, waste management and planning as well as other areas.

Greater Manchester has already a combined authority responsible for economic development, business support, policing and skills as part of the "Northern Powerhouse". With a directly elected mayor for the whole Greater Manchester metropolis, there is substantial opportunity to influence the regional environment. Opportunities for changing the environmental agenda will go beyond England with environmental quality as a leading contender for greater control. The challenge for countries, cities and towns picking up new environmental responsibilities will be to find the funding to achieve their environmental ambitions, particularly taking account the conflicting pressures and priorities.

Looking forward, the economy and international relations are likely to continue to dominate government policy. As could be seen in the run-up to the referendum on membership of the EU, the environment was not a high public or political priority. This was despite strenuous efforts by some professional bodies including the Society for the Environment (SocEnv) and the IEMA to raise awareness of environmental issues including regulation and standards in the debate. While Friends of the Earth acknowledged the beneficial impact of EU membership on the protection of the environment, many bodies kept a more neutral position possibly due to fear of antagonising their members (Burns 2015). The beneficial impact of the EU waste, water, air quality and habitat regulation unfortunately was seen by many as a loss of sovereignty and excess bureaucracy. At least some political leaders are committed to continuing to contribute to and deliver EU environmental policies despite Brexit.

The Scottish Environment Secretary Roseanna Cunningham said to environmental organisations in a letter in July 2016 (The Scottish Government 2016):

> We will be doing our absolute utmost to protect our position as climate change leaders, to continue to play the role that we have been playing in contributing to EU-wide environmental policies, and to make sure that we maintain, protect and advance our own environment.

The focus of policy making is likely to continue to be on the implementation of cuts in the staff and other resources in Westminster and devolved administrations. DEFRA's strategy to 2020 published in January 2016 says "a strong economy needs a healthy environment, and a healthy environment needs a strong economy" (Department for Environment, Food and Rural Affairs 2016). Its objectives are:

1. a cleaner, healthier environment, benefiting people and the economy
2. a world-leading food and farming industry
3. a thriving rural economy, contributing to national prosperity and well-being
4. a nation better protected against floods, animal and plant diseases and other hazards with strong response and recovery capabilities
5. excellent delivery on time and to budget and with outstanding value for money
6. an organisation continually striving to be the best, focused on outcomes and constantly challenging itself
7. an inclusive, professional workforce where leaders recognise the contribution of people and build capability to deliver better outcomes

What is very unclear is how, against continuing austerity cuts, it will actually deliver the first four objectives. With the reduction of its skills base, this is going to fall on other hard-pressed agencies, authorities and organisations. What we will see is a tighter cost verses benefit approach to the need for or any tightening of environmental regulation. While the strategy aspires to better conditions for food, farming and rural businesses, environment protected, it also commits to less red tape for UK businesses overall.

On the regulation of energy, we will continue to see the decarbonising of the electricity sector including the closure of all the existing coal-fired power stations. While there will be support and possible government subsidies for new gas build, fracking and new nuclear build, the reduction in subsidies for renewable energy will continue. The UK government focus is on safe, secure and future-proof supplies of energy available to all (Department of Energy and Climate Change 2016b).

Due to the pushback against regulation and red tape both in the UK and across the EU and the pressures on governments to address other issues such as mass migration and the economic outlook it is unlikely that we will see much in the way of new or changes to existing environmental regulation in the next few years. An increase in the demand for deregulation as a result of changes in the relationship between the UK and the rest of the EU is more likely to focus on individual freedoms. However, as has been found in the past, we cannot always predict the future.

It is likely that an unwanted or unanticipated environmental issue or impact will arise. Possible examples already with us include waste plastics in the marine environment including micro-beads, carbon particles in the air or our failure to adequately reduce greenhouse gas emissions. Globally and locally, we are seeing serious loss of soil and increased flood risk due to land use and development. The international movement of electronic goods is both potentially causing a shortage of critical raw materials and contamination. Without us all taking full individual responsibility, some forms of control will be required to prevent and minimise unwanted releases and manage our wastes better. If we are to achieve sustainable development, environmental regulations will continue to be with us and benefit us.

1.7 Conclusion

Environmental regulations historically have been introduced in response to specific threats to the environment and as a result of the activities of particular industries. They have developed to encompass activities impacting the air, land and water both in the short term, for example in air and water quality, and in the longer term addressing climate change. Increasingly, regulations have been implemented to deliver international issues and agreements and have been developed at a European level, predominately through the EU. This has required international collaboration and achieved a high level of consistency by industry and regulators.

The style of regulations has evolved, placing more responsibility on businesses to identify, justify and use the best techniques. Increasing use has been made of management standards and voluntary agreements with a regulatory/enforcement backup. There are now regulations derived and delivered at local, national and international levels.

The wide range of regulatory measures, guidance and standards now applying to businesses has given rise to challenges, reviews and repeal of some legislation. The separation of the UK from the EU will result in uncertainty as to which European-derived legislation will continue to apply and how. It will also give an opportunity to review and revise environmental policies and regulations. The challenge will be to keep the best of existing UK and EU environmental regulations, influencing and implementing future regulations in areas where they are deficient.

References

Andrew Gouldson, J.M. 2013. *Regulatory Realities: The Implementation and Impact of Industrial Environmental Regulation*. s.l.: Routledge.

Brimblecombe, P. 2002. The Great London Smog—Effects and Repercussions. In *The Clean Air Revolution*, ed. T. Williams and L. Murley, 10–12. Brighton: National Society for Clean Air and Environmental Protection.

Burns, C. 2015. *The Implications for UK Environmental Policy of a Vote to Exit the EU.* https://www.foe.co.uk/sites/default/files/downloads/eu-referendum-environment-81600.pdf. Accessed 27 July 2016.

Christine Berry, S.D. 2015. *Threat to Democracy, the Impact of 'Better Regulation' in the UK.* London: New Economics Forum.

Commission to the Council, the European Parliament, the Economic and Social Committee and the Committee of the Regions. 2001. *Sixth Environment Action Programme of the European Community.* Environment 2010: Our future, Our choice, Com 2001/31. s.l.: s.n.

Council and the Representatives of the Governments of the Member States. 1993. *Fifth European Community Environment Programme: Towards Sustainability*. s.l.: s.n.

Davidson, N. 2006. *Davidson Review Implementation of EU Legislation*. s.l.: Better Regulation Executive Cabinet Office.

Department for Environment, Food and Rural Affairs. 2012. *Environmental Permitting. General Guidance Manual on Policy and Procedures for A2 and B Installations*. https://www.gov.uk/government/publications/environmental-permitting-general-guidance-manual-on-policy-and-procedures-for-a2-and-b-installations. Accessed 28 Mar 2017.

Department for Business, Innovation and Skills. 2011a. *One-in, One-out: Statement of New Regulation*. London: HM Government.

———. 2011b. *Red Tape Challenge Kicks Off*. s.l.: HM Government.

———. 2016. *Cutting Red Tape Review of the Waste and Recycling Sector*. s.l.: HM Government.

Department for Environment, Food and Rural Affairs. 2011. *Environment Regulations Under the Red Tape Spotlight*. https://www.gov.uk/government/news/environment-regulations-under-the-red-tape-spotlight

———. 2012. *Red Tape Challenge—Environment Theme Proposals*. London: Department for Environment, Food and Rural Affairs.

———. 2015. *Emerging Findings from Defra's Regulation Assessment*. s.l.: s.n.

———. 2016. *Creating a Great Place for Living: Defra's Strategy to 2020*. https://www.gov.uk/government/uploads/system/uploads/attachment_data/file/501709/defra-strategy-160219.pdf. Accessed 27 July 2016.

———. 2017. *The UK's Implementation of the Pollutant Release and Transfer Register (PRTR) Under the Kiev Protocol—Call for Views*. https://consult.defra.gov.uk/atmosphere-local-environment-team/copy-of-protocol-on-prtrs/. Accessed 20 Mar 2017.

Department for Environment, Food and Rural Affairs. 2014. *DEFRA Better FOR BUSINESS a Strategic Reform Plan for Defra's Regulations*. s.l.: s.n.

Department of the Environment Northern Ireland. 2015. *Environmental Governance in Northern Ireland*. s.l.: s.n.

Department of Agriculture, Environment and Rural Affairs. 2016. *Northern Ireland Environment Agency*. https://www.daera-ni.gov.uk/northern-ireland-environment-agency. Accessed 25 July 2016.

Department of Energy and Climate Change. 2016a. *Corporate Report Single Departmental Plan: 2015–2020*. https://www.gov.uk/government/publications/decc-single-departmental-plan-2015-to-2020/single-departmental-plan-2015-to-2020. Accessed 27 July 2016.

———. 2016b. *The Onshore Hydraulic Fracturing (Protected Areas) Regulations (SI 2016/384)*. s.l.: s.n.
EEF. 2015. *Green Tape: Manufacturers' Views of Progress on Defra's Regulatory Reform Agenda*. s.l.: EEF.
Ends. 1992. *Integrated Pollution Control and the Organic Chemicals Industry*. http://www.endsreport.com/article/397/integrated-pollution-control-and-the-organic-chemicals-industry. Accessed 28 July 2016.
———. 1996. *Big Savings at Allied Colloids Confirm "3Es" Success*. http://www.endsreport.com/article/3155/big-savings-at-allied-colloids-confirm-3es-success. Accessed 26 July 2016.
———. 1998. *Emission Constraints Rock the Boat for Power Industry Shake-Up*. http://www.endsreport.com/article/5117/emission-constraints-rock-the-boat-for-power-industry-shake-up. Accessed 27 July 2016.
———. 1999. *Pollution Inventory Redraws the Map for Environmental Information*. http://www.endsreport.com/article/5502/pollution-inventory-redraws-the-map-for-environmental-information. Accessed 28 July 2016.
———. 2000. *Onyx Rebuilds Internal Audit Culture After Record Waste Fines*. http://www.endsreport.com/article/6232/. Accessed 26 July 2016.
———. 2001. *Agency Proposes New Shift to Self-monitoring by Industry*. http://www.endsreport.com/article/7371. Accessed 26 July 2016.
———. 2004. *Sevalco: Driving a Coach and Horses Through Self-regulation*. http://www.endsreport.com/article/13484/. Accessed 26 July 2016.
———. 2006. *Risk-based Regulation: Planning for Compliance*. http://www.endsreport.com/article/15640/risk-based-regulation-planning-for-compliance. Accessed 26 July 2016.
———. 2010a. Coalition Pledges 10% Cut in Government's Emissions. *Ends*, May 20.
———. 2010b. *St Regis Found Guilty of Falsifying Effluent Data*. http://www.endsreport.com/article/25838/st-regis-found-guilty-of-falsifying-effluent-data. Accessed 26 July 2016.
———. 2012. *Green Certified Standards*. London: Ends.
———. 2016. *Ends Report 498 Brexit Overview*. http://www.endsreport.com/downloads/53357.pdf. Accessed 27 July 2016.
Environment Agency. 1997. *Operator and Pollution Risk Appraisal*. Bristol: Environment Agency.
———. 2002. *Environmental Protection OPRA System. Proposed Charging Scheme*. Bristol: Environment Agency.

———. 2003. *IPPC Regulatory Guidance Series No 7: Criteria for Determining Whether an Installation Can Be Classified as Low Impact*. Bristol: Environment Agency.

———. 2004. *Delivering for the Environment: A 21st Century Approach to Regulation*. Bristol: Environment Agency.

———. 2014a. *EPR Assurance Trial*. Bristol: Environment Agency.

———. 2014b. *Regulating for People, the Environment and Growth*. https://www.gov.uk/government/uploads/system/uploads/attachment_data/file/312703/LIT_9902.pdf. Accessed 26 July 2016.

———. 2015. *Regulatory Guidance Series, No RGN 2 Understanding the Meaning of Regulated Facility*. https://www.gov.uk/government/uploads/system/uploads/attachment_data/file/435475/LIT_6529.pdf. Accessed 14 Nov 2016.

———. 2016. *Environment Agency*. https://www.gov.uk/government/organisations/environment-agency/about. Accessed 25 July 2016.

Environment Agency R&D Technical Report P6-017/2/TR. 2003. *Environmental Management Systems & Operator Performance at Sites Regulated under Integrated Pollution Control*. Bristol: Environment Agency.

European Commission. 2012. *The Lisbon Treaty*. Official Journal of the European Union C 326, October 26.

———. 2016. *Interinstitutional Agreement between the European Parliament, the Council of the European Union and the European Commission on Better Law-making*. s.l.: s.n.

European Community OJ C112/1. 1973. *Declaration of the Council of the European Communities and of the Representatives of the Governments of the Member States Meeting in the Council of 22 November 1973 on the Programme of Action of the European Communities on the Environment*. s.l.:s.n.

European Council. 1996. *Directive 96/61/EC on Integrated Pollution Prevention and Control*. s.l.:s.n.

———. 2016. *Guide to the Ordinary Legislative Procedure*.

European Parliament and Council. 2008. *Directive 2008/1/EC Concerning Integrated Pollution Prevention and Control*. s.l.:s.n.

———. 2010. Directive 2010/75/EU on Industrial Emissions (Integrated Pollution Prevention and Control. 24 November).

European Union. 2014. *General Union Environment Action Programme to 2020*. Luxembourg: Publications Office of the European Union.

Falkner, Gerda, M. Hartlapp, S. Leiber, and O. Treib. 2004. Non-Compliance with EU Directives in the Member States: Opposition Through the Backdoor. *West European Politics* 27 (3): 452–473.

Federation of Small Businesses. 2011. *The Burden of Regulation*. http://www.fsb.org.uk/LegacySitePath/policy/images/the%20burden%20of%20regulation%20-%20may%202011.pdf. Accessed 26 July 2016.

Frankfurt School—UNEP Collaborating Centre for Climate & Sustainable Energy Finance. 2016. *Global Trends in Renewable Energy Investment 2016*. Frankfurt: Frankfurt School of Finance & Management gGmbH.

GOV.UK 2014. *Prime Minister Announces Government Exceeds Its Target to Identify 3,000 Regulations to be Amended or Scrapped*. http://webarchive.nationalarchives.gov.uk/20150522175321/http:/redtapechallenge.cabinetoffice.gov.uk/themehome/pm-speech-2/

Hansard. 1972. Dumping of Cyanide (Nuneaton). *House of Commons Debate* 831: 1667–1672.

Hey, C. 2005. EU Environmental Policies: A Short History of the Policy Strategies. In *EU Environmental Policy Handbook*, ed. S. Scheuer, 17–30. s.l.: European Environment Bureau.

Hjerp, P., I. Humeyer, M. Pallemaerts, and A. Farmer. 2010. *The Impact of Better Regulation on EU Environmental Policy Under the Sixth Environment Action Programme*. s.l.: The Brussels Institute for Environmental Management (Bruxelles Environnement/Leefmilieu Brussel), IEEP.

House of Commons Environment, Food and Rural Affairs Committee. 2006. *The Environment Agency Seventh Report of Session 2005–6*. London: House of Commons.

———. 2016. *Future Flood Prevention, Second Report of Session 2016–17*. http://www.publications.parliament.uk/pa/cm201617/cmselect/cmenvfru/115/115.pdf. Accessed 21 Nov 2016.

House of Commons Transport Committee. 2016. *Volkswagen Emissions Scandal and Vehicle Type Approval*. London: House of Commons.

International Council of Chemical Associations. 2016. *Responsible Care*. https://www.icca-chem.org/responsible-care. Accessed 27 July 2016.

Kaminski, I. 2016. *DEFRA Urged EA to Deny Its Preferred Chief Executive Candidate*. http://www.endsreport.com/article/51167/defra-urged-ea-to-deny-its-preferred-chief-executive-candidate. Accessed 27 July 2016.

Lévêque, François. 1998. *Voluntary Approaches, Environmental Policy Research Briefs Number 1*. https://www.ucd.ie/t4cms/pb-ep-01.pdf. Accessed 27 July 2016.

McCarthy, D., and P. Morling. 2015. *Using Regulation as a Last Resort: Assessing the Performance of Voluntary Approaches*. Sandy, Bedfordshire: Royal Society for the Protection of Birds.

Meadows, Donella H., Dennis L. Meadows, Jørgen Randers, and William W. Behrens III. 1972. *The Limits to Growth*. s.l.: Club of Rome.

National Driver Offender Retraining Scheme. 2016. *Trends and Statistics*. s.l.: s.n.

Natural Resources Wales. 2016. *Business Plan 2016/17*. https://naturalresources.wales/media/677799/business-plan-document-2016-17.pdf. Accessed 14 Nov 2016.

NSCA Commission on Industrial Regulation and Sustainable Development. 2001. *Smarter Regulation*. Brighton: NSCA.

O'Riordan, T., and B. Wynne. 1987. Regulating Environmental Risks: A Comparative Perspective. In *Insuring and Managing Hazardous Risks: From Seveso to Bhopal and Beyond*, ed. P.R. Kleindorfer and H.C. Kunreuther, 389–410. Berlin: Springer.

Perring, R. 2016. *REVEALED: The EU's Top Ten Pointless Decisions the UK Can Now Get Rid of*. London: Daily Express Newspaper.

Planet and Prosperity Ltd. 2013. *New Opportunities to Improve Environmental Compliance Outcomes Using Certified EMss*. Edinburgh: Sniffer.

Rezessy, Silvia, P. Bertoldi, and A. Persson. 2005. *Are Voluntary Agreements an Effective Energy Policy Instrument? Insights and Experiences from Europe*. s.l.: European Commission Joint Research Centre.

Royal Commission on Environmental Pollution. 1976. *Fifth Report: Air Pollution Control—An Integrated Approach* [Cmnd. 6371]. London: HMSO.

———. 1984. *Tenth Report: Tackling Pollution—Experience and Prospects* [Cmnd, 1419]. London: HMSO.

———. 1985. *Eleventh Report Managing Waste: The Duty of Care*. London: HMSO.

———. 1988. *Twelfth Report Best Practicable Environmental Option*. London: HMSO.

———. 1993. *Seventeenth Report Incineration of Waste*. London: HMSO.

Schwab, K. 2015. *The Global Competitiveness Report 2015–2016*. Geneva: World Economic Forum.

SEPA. 2016. *Annual Operating Plan 2016–2017*. s.l.: s.n.

Statutory Instruments. 2015. *Environment Protection, England the Single Use Carrier Bags Charges (England) Order SI 2015 No.776*. s.l.:s.n.

Survey, B.A. 2015. *The Ozone Hole*. https://www.bas.ac.uk/data/our-data/publication/the-ozone-layer/

The Commission of the European Communities. 1999. *Commission Recommendation on the Reduction of CO2 Emissions from Passenger Cars (1999/125/EC)*. s.l.: Official Journal of the European Communities.

The Scottish Government. 2016. *EU Environment Pledge.*
UK Parliament. 2013. *Environmental Permitting (England and Wales) (Amendment) Regulations 2013 (SI 2013 No. 390)*. London: The Stationary Office.
———. 2015. *The Single Use Carrier Bags Charges (England) Order SI 776.* s.l.: s.n.
www.remas.info. 2006. *Remas Linking Environmental management and Performance.*

Martin Bigg is Professor of Environmental Technologies Innovation at the University of the West of England (UWE), Bristol, the UK, and Chair of Council and Board Member of the Institute for Environmental Management and Assessment. He leads and supports environmental initiatives with governments, businesses, local authorities and communities. As Director of the Environmental Technologies Innovation Network (iNet), he led a European programme to support innovative enterprising businesses. He was Chair of the Bristol Green Capital Partnership, which secured the title of European Green Capital 2015 for the City of Bristol. Previously, he was the Head of Industry and Waste Regulation at the Environment Agency, England and Wales.

2

Environmental Regulation and Growth: Impact on Sustainable Economic Growth

Jonathan Fisher

2.1 Introduction

This chapter examines the economic impact of environmental regulations. Section 2.2 presents estimates of the costs and benefits of environmental regulations and their impact on the growing environmental goods and services sector (EGSS) in the UK.

Section 2.3 reviews the available evidence on the impacts of environmental regulations on economic growth, innovation and technical change as well as impacts on competitiveness and any movement of businesses to less well-regulated pollution havens. This includes an examination of the Porter Hypothesis (PH) that environmental regulations can trigger greater innovation that may partially or more than fully offset the compliance costs. Section 2.4 then sets out principles for how better environmental regulation can improve its impacts on sustainable economic growth and illustrates how the Water Framework Directive (WFD) is a good example of the application of these principles in practice. Section

J. Fisher (✉)
Jonathan Fisher Environmental Economics, Hatfield, UK

J. Foreman (ed.), *Developments in Environmental Regulation*, Palgrave Studies in Environmental Policy and Regulation, DOI 10.1007/978-3-319-61937-8_2

2.5 then examines calls for greater certainty regarding future environmental regulations, whereas in fact there should be calls for less uncertainty. This section then suggests how this could be achieved.

Section 2.6 reviews current and recent political perspectives regarding developments in environmental regulations across the EU and shows how the UK has successfully positively managed to influence such developments so that the EU environmental regulations now incorporate many of these principles to improve their impacts on economic growth. See section "Issues Regarding Brexit" then examines the implications of Brexit for UK environmental regulations.

Finally, Section 2.7 sets out some best practice principles to improve the impacts of environmental regulation on sustainable economic growth, innovation and technical change.

2.2 Costs and Benefits of Environmental Regulation

Net Costs to Business

Direct costs of regulations to business include (annualised) investment and operating costs (e.g., in pollution control equipment) and administrative costs (e.g., form filling); the latter amount to about £808 m p.a. (14% of the direct costs to business) (Defra 2015).

Department for Food, Environment and Rural Affairs (Defra (2015: 1)) reviewed the available estimates for these costs—drawing largely on findings of Impact Assessments of specific regulations plus other studies and cross-checking the various data. They found that the gross direct costs to business of environmental regulations[1] in 2012 were about £6 bn p.a. (in 2012 prices). Figure 2.1 shows these costs for Defra's main policy areas. Figure 2.2 shows which industry sectors (e.g., water, manufacturing, agriculture, fisheries and food) incurred these costs. Defra (2015: 10) shows that environmental regulation's gross costs to all businesses account for just 0.16% of turnover of all businesses and only 0.2% of turnover of manufacturing businesses. These costs of environmental regulations are

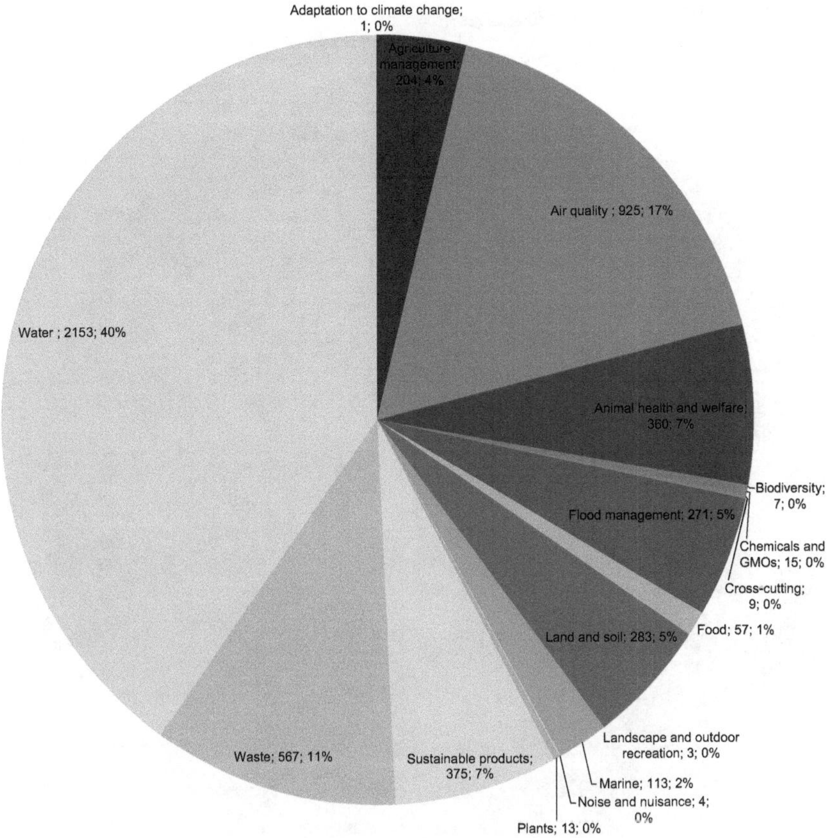

Fig. 2.1 Direct costs to business of Defra's regulations by policy area, 2012 (£m, %). Source: Defra (2015). Emerging Findings from Defra's Regulation Assessment, First update covering 2012

lower than other regulations. For water industry and waste, the gross costs of environmental regulations account for 12% and 2.5%, of their turnover, respectively. For agriculture, fish and forestry, the costs are 3% of turnover.

Defra also estimated that these regulations had direct benefits to businesses (e.g., more efficient use of resources and energy) amounting to about £2 bn giving net costs to business that are about one-third lower at about £4 bn p.a.

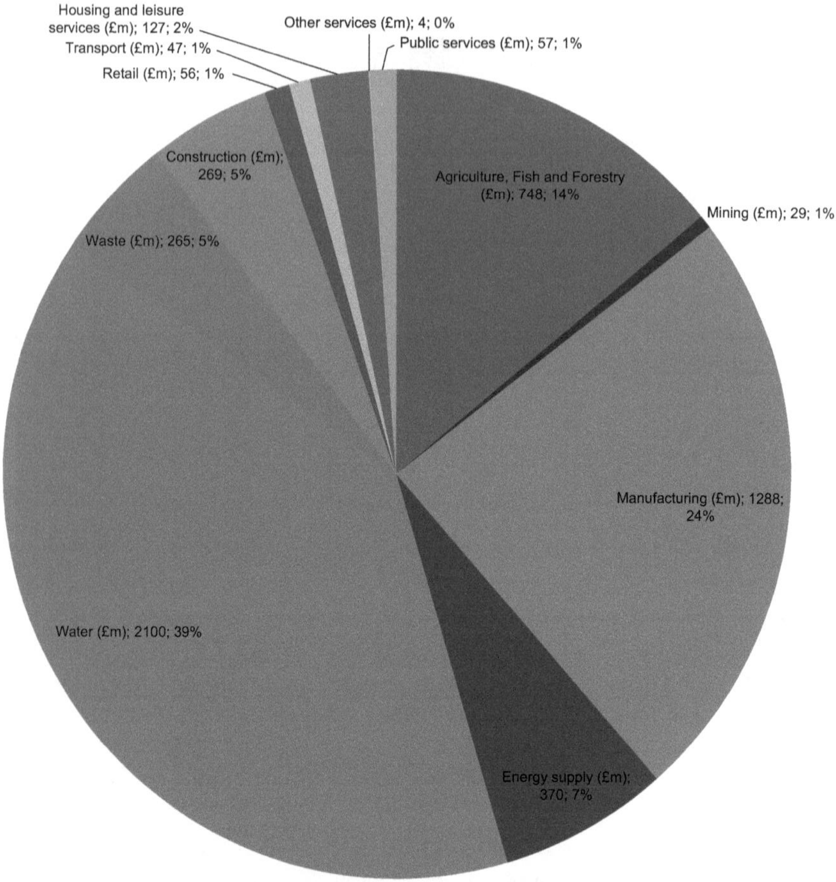

Fig. 2.2 Direct costs to business of Defra's regulations by industry sector, 2012 (£m, %). Source: Defra (2015). Emerging Findings from Defra's Regulation Assessment, First update covering 2012

The European Commission (2015) reports on environmental expenditures for the period 1995–2012 show that these expenditures declined in absolute real terms between 1996 and 2012 and have fallen to become about 2–3% of value added in 2012.

Defra (2015: 5) reports that the number of domestic UK environmental regulations in 2012 increased by one while the number of EU regulations declined by six. Likewise, the Economist (2016a: 24) states that "the EU is proposing far fewer rules now. The European Commission's better regulation agenda limits new regulations and even withdraws existing ones. It is ironic that Britain should consider Brexit just when the EU has come round to a more competitive, less intrusive approach" (see Sect. 2.6).

Benefits of UK Environmental Regulations

Environmental regulations yield significant benefits in terms, for example, of improving people's health and welfare and reducing environmental damage caused by pollution as well as enhancing ecosystems and the natural capital on which people's welfare and future prosperity depends (see Defra 2011, 2014; Natural Capital Committee 2014).

As part of their analysis of costs reported the section "Net Costs to Business" above, Defra (2015: 1) also estimate that the monetised benefits of environmental regulations amount to about £10 bn p.a. Accordingly, they estimate that the ratio of their estimates of the monetised benefits to net costs (BCR) of environmental regulations is about 3:1. In addition, there are non-monetised benefits arising from, for example, improvements in natural habitats and ecosystems.

Consequently, the Aldersgate Group (2011) argue that the focus should be on increasing the effectiveness and efficiency with which environmental regulations address environmental pressures and achieve these benefits better through better regulation measures—rather than headline-grabbing initiatives to cut the regulations in any "bonfire of the regulations".

Environmental Goods and Services Sector (EGSS) in EU and UK

Eurostat (2016) reports that the EU market for environmental goods and services has grown by about 50% between 2003 and 2013. The UK has a significant share of this international market. The EGSS is not conventionally a sector as such in the national economic accounts. Nevertheless, the Office for National Statistics (ONS) has derived estimates for the

EGSS through satellite accounts and using the commonly accepted definition for the EGSS developed by Eurostat and adopted by the System of Environmental Economic Accounting.

The EGSS contributes significantly to the UK economy. ONS (2015) reports that the UK EGSS in 2012 had an output of £55 bn and a Gross Value Added (GVA) of £26 bn (or 1.6% of GDP). This exceeds the GVA of the agricultural sector (at £10 bn in 2012). EGSS employed 357,200 people in full time employment in 2012. Between 2010–2012, its output, GVA and employment grew by 9.1%, 1.5% and 5.3%, respectively.

The Dutch and German Governments assist businesses to pursue export opportunities for this sector. The environmental regulator in Scotland (the Scottish Environmental Protection Agency (SEPA)) participates in the European Commission's Environmental Technology Verification (ETV) pilot programme which provides independent verification of the performance and environmental benefits of a new technology to accelerate its market entry.

2.3 Economic Impacts of Environmental Regulations

Impact of Environmental Regulation on Sustainable Economic Growth

Martin Bigg, in Chapter 1, reports that UK environmental regulators have high-level objectives to support growth. Thus, the EA has an objective to "support sustainable growth". SEPA has an objective to "protect and improve the environment in ways that, as far as possible, also help create health and wellbeing benefits and sustainable economic growth". The Northern Ireland EA has an objective to "support a sustainable economy". One of Defra's objectives is "A cleaner, healthier environment benefitting people and the economy". Therefore, this chapter examines how environmental regulations could affect economic growth and then reviews the evidence on this subject.

Environmental regulations could affect economic growth through their impacts on the following factors in Her Majesty (HM) Treasury (2015: 11)'s framework for raising productivity:

1. *Business investment for the long term:* On the one hand, environmental regulations increase investments in environmental protection measures and economic activity in the EGSS (see the section "Environmental Goods and Services Sector (EGSS) in EU and UK") and they might bring underused resources into use thereby increasing economic activity. However, they entail net costs to business (see the section "Net Costs to Business"), which Sato (2014) states take resources that businesses could otherwise use to implement more productive investments. Moreover, any delays in securing permits can discourage business investments.
2. *Enhancing skills and human capital:* HM Government (2012) and Aldersgate Group (2012b) stress the importance of enhancing skills to achieve more efficiently environmental improvements, sustainable growth and a transition to a green economy. Natural Environment Research Council (NERC) (2012) identified 15 critical skill gaps in the environmental sector.

   The prime minister in her speech to the Confederation of British Industry (CBI) conference (May (2016)) said "we are not strong enough in STEM subjects, and our technical education isn't good enough". Policy on skills is a devolved matter. In England, the policy is demand led by the needs of business. But Defra (2011: 38) reports that businesses are uncertain about future green skills needs and the skills requirements and opportunities for progress in achieving a successful green economy. The Aldersgate Group (2012b: 4) sees a vital role for government to intervene to ensure that education and training is designed for future needs and not merely to remedy current shortages. The coalition government (in HM Government (2012: 9)) concluded that new "skills for a green economy" grouping of Sector Skills Councils could help business understand changing skills requirements. Other proposed actions included improving the quality of information advice and guidance available on careers in a green economy together with information on the skills linked to the green economy that will be needed in the future. They also recommended improving the quality of skills provision in the further education system and raising awareness and understanding of the green economy to support lifelong learning among the workforce. Since then, there appears to have been little action on these matters. HM Government (2017) emphasised the importance of enhancing skills, especially technical skills. This is the second pillar in their Green Paper setting out their proposed Industrial Strategy, which includes boosting Science, Technology, Engineering and Maths skills and raising skill levels in lagging areas. However, as yet (January 2017), the current government does not appear to address environmental skills needs.

3. *Improving water and flood defence infrastructure* is an important element of the third pillar of the government's Green Paper on Industrial Strategy. Ensuring sustainable provision of water resources, waste water management and flood risk management supports and sustains economic activities dependant on such resources. Thus, the EA's regulation of abstractions and work on Water Resource Management Plans and abstraction incentive mechanisms and abstraction trading can enable economic development to proceed in water-stressed catchments while still ensuring sustainable water resources and environmental safeguards. For example, Pepsico is significantly reducing their water consumption of their agricultural and agro-industrial operations (e.g., crisp manufacture) (see Pepsico 2010). They collaborated with the EA to increase the water efficiency of Pepsico's farm businesses, cutting water consumption by 30% and enabling it to grow sustainably in a water-stressed catchment. Similarly, the Natural Capital Committee (2014: 56) reports that Walkers Crisps increased the water efficiency of their manufacturing processes which saved £630 k p.a. and enabled them to continue operating in a seriously water-stressed area.
4. *Enhancing ideas, knowledge and development and adoption of innovations and technical change:* Angus et al (2013: 8) and Sato (2014) report that environmental regulation has been a positive driver of innovation, especially in achieving more effective and efficient environmental protection and may increase overall innovation by large firms, especially in the long run when it leads to changes in corporate strategies to enhance innovation. But Angus et al (2013: 9) also reports that R&D driven by environmental regulations has displaced other R&D especially for small firms.

5. *Flexible, fair markets, openness and competition:* Environmental regulations can provide an even-playing field for all businesses and hence not only prompt them to adopt efficient control measures but also provide a clear basis for the growth of the EGSS (see the section " Environmental Goods and Services Sector (EGSS) in EU and UK"). The EA's work on controlling waste crime can prevent illegal operators undercutting otherwise worthwhile, efficient and viable firms. This can provide a good basis for environmentally preferable firms acting legally as well as reducing the illegal operators' significant environmental damages.
   However, there are risks that big companies could dominate and this could lead to regulatory capture and inhibit the entry of entrepreneurial (small) companies offering environmentally and economically attractive alternative products. For example, the big six waste management companies have tended to dominate waste management policies and strategies. They have pressed county councils to commit to large incinerators and waste management facilities on 25-year contracts under Private Finance Initiative (PFI) funding schemes,[2] which restrict the entry into the waste market of entrepreneurial companies offering smaller scale and more environmentally and economically attractive alternative waste management options. Other European countries seem to promote more positively the adoption of such options and moves to achieve a more circular economy.
   Moreover, tradeable permit systems that grant (grandfather) rights to existing firms can adversely affect growth by preventing new firms (especially dynamic small firms) from entering the market.

Impacts on International Competitiveness and Pollution Havens

The impact on international competitiveness is the combination of the effects on production costs, productivity, innovation and technical change. Angus et al (2013) reports some studies that show that environmental regulations have impaired industries' competitiveness and led to pollution-intensive industries moving to less regulated jurisdictions. Ederington et al. (2005) found that the impact of regulation on competitiveness is negative but small. They also show that those industries with the largest pollution abatement costs also happen to be the least geographically mobile or footloose. Therefore, the concerns may not be so much that the industries move investments overseas but rather that UK plants close in the face of strong international competition especially where there is global overproduction—as for steel.

However, Angus et al (2013) also reports that several studies found no evidence that environmental regulations have negatively affected trade and competitiveness and investment moving to countries with lax environmentalregulations—see also Sato (2014) and Mani and Wheeler (1998). Environmental regulation's costs are a small percentage of businesses' turnover (see the section "Net Costs to Business"). Other costs (e.g., labour, energy) and other factors such as access to a big market, well-defined property rights, good governance and a good regulatory environment are much more important in determining foreign direct investment.

Moreover, investments in new plant concern the long term. Therefore, the key comparison for business decision-making is not between current environmental regulations and those currently in other countries but with the future regulations in these other countries over the length of the plant in question. The UK is in a strong position to help these countries develop their policies and programmes to address these problems, and UK firms in the EGSS are well placed to export technologies, products and systems to help them solve these problems.

Overview of Evidence

The available evidence on all these matters is contested and not clear cut. Specific, partial or anecdotal studies claiming that environmental regulations have significant positive or negative economic impacts need to be treated with considerable caution. For example:

A. There are often claims that environmental regulations increase growth by enabling the growth of the UK EGSS (see the section "Environmental Goods and Services Sector (EGSS) in EU and UK"). Such growth is important and the environmental regulations need to be designed and implemented to maximise the beneficial impacts on this important UK business sector. But these outputs are paid for by other business' costs (as shown in the section "Net Costs to Business"). So, beneficial growth of the UK EGSS offset impacts on growth of such costs rather than necessarily being overall net positive impacts on growth.

B. Conversely, one needs to be wary of claims of environmental regulations imposing very high costs and significant adverse economic impacts. Morgenstern (2016: 8)'s retrospective analyses of 34 case studies of US environmental regulations found that there is a tendency to overstate the costs and benefits of regulations in the analyses done before the regulations were issued. Similarly, Fisher (2008) found that water companies' estimates of the costs of environmental measures in their draft business plans were about 40% higher than their costs in the final business plans once the Water Services Regulatory Authority (OFWAT) and the EA had scrutinised their estimates to remove over-estimation and gold plating. OFWAT and the EA then sought more efficient alternatives and refined the requirements accordingly. Morgenstern (2016: 10) suggests that "One thing that the agencies could do is to build a plan of retrospective analysis into the regulation at the time it's promulgated… Obviously you can't do it for all rules—especially in times of tight budgets. You should be selective". The government now requires post-implementation reviews of regulations (see Department for Business, Innovation and Skills (2016)). This is largely a qualitative review that examines the following questions:

   a. Did the policy or regulation achieve its objectives?
   b. Did it have any unintended consequences?
   c. Were there any opportunities to reduce burdens on business?
   d. How does the UK's implementation compare with that in other EU member states in terms of costs to business?

It could also be worthwhile carrying out some empirical analysis comparing the actual costs to business with the ex ante estimates in the Impact Assessment. But requiring this for all regulations would impose an excessive burden for regulators who need to focus efforts on improving the design and implementation of regulations to improve their impacts (see Sect. 2.4). So perhaps such plans for retrospective empirical analysis should just be required for any regulation for which the Impact Assessment showed the estimated costs would exceed a certain threshold—perhaps 2% or 5% of the regulated sectors' turnover. Moreover, the lessons from such retrospective analyses need to be taken into account in designing and implementing future regulations and estimating their costs and benefits. Their findings should be incorporated into databases of unit costs of control measures used in future Impact Assessments.

C. There has been much debate about the Porter Hypothesis (PH), though Ambec et al. (2016) state that there is "oftentimes a misunderstanding of what the PH does and does not say". The "PH" expounded in the seminal work by Porter and Linde (1995b: 98) states that "properly designed environmental standards *(can)* trigger innovation that may partially or more than fully offset the costs of complying with them". Ambec et al. (2016: 2) add that environmental standards can do so "in some instances". Porter and Linde do not actually say or conclude that environmental regulations will (always) reduce costs and enhance competitiveness. Porter and Linde (1995b: 100) actually say: "We readily admit that innovation cannot always completely offset the cost of compliance, especially in the short term before learning can reduce the cost of innovation-based solutions". Similarly, Sato (2014: 4) conclude that "green innovations developed to reduce the cost of environmental regulations do not seem to increase firms' profits enough to fully offset the private costs of regulation". The extent of such offsetting depends on the potential for environmental regulations to lead to increased innovation by businesses, which is probable since there are many untapped worthwhile innovations and technical changes that businesses need an external spur from regulations to consider and implement.

Porter and Linde (1995a: 130) state: "Certainly, misguided regulatory approaches have imposed a heavy burden on companies" and (in Porter and Linde (1995b: 98)) that "these costs are far higher than they need to be". They also say that successful visionary companies tend to have a better environmental record.

Conclusions to take from Porter's work are in fact the following:

a. The need for properly designed and implemented regulations and for regulators with strong technical capabilities to work collaboratively with businesses.
b. Businesses need to devote their efforts to respond positively and innovatively to environmental concerns and pressures rather than being adversarial and litigious.
c. Successful visionary businesses with dynamic corporate strategies have better environmental and economic records.
d. There is a need to link environmental regulations with industrial policies to promote innovation and technical changes in businesses and their corporate strategies. This should include dissemination of information and training about latest best practice techniques. Technically capable environmental regulators and trade associations can play an important role in informing businesses about efficient best practice ways of tackling specific environmental problems.
e. Growth is driven by improving total factor productivity, which includes all factors of production such as capital, labour and raw materials. There is a need to incorporate into this positively natural resources and the need for improved resource productivity that Porter and Linde (1995b) highlight and advocate.
f. Greater environmental benefits and improvements in natural capital are needed to sustain economic activity and growth (see the section "Benefits of UK Environmental Regulations").

Summary on Impacts of Environmental Regulations on Economic Growth

Nevertheless, the following conclusions can clearly be drawn from the available evidence:

1. Angus et al (2013: 15) conclude that "The weight of evidence suggests that there is no <u>significant</u> economic impact of environmental regulation". Defra (2015: 10) show that environmental regulation's gross costs to all businesses account for just 0.16% of their total turnover

and only 0.2% of turnover of manufacturing businesses. For the most polluting sectors of water and waste, the gross of environmental regulations account for 12% and 2.7% of their turnover, respectively. For agriculture, fisheries and forestry, they account for 3% of turnover. The net costs to business of environmental regulations are about one-third lower when account is taken of the financial benefits of environmental regulations to business (see the section "Net Costs to Business").
2. What is clear is that better environmental regulation can significantly enhance their benefits and improve their impacts on economic growth and their positive impacts on the UK EGSS. The next section explores this aspect. This builds on the better regulation initiatives discussed in the other chapters of this book.

2.4 How to Improve Environmental Regulation's Economic Impacts

This section set out principles for regulatory design that will improve their economic impacts. It draws on recommendations in the literature (e.g., Porter and Linde 1995a: 124) and also best practice guidance for regulations (e.g., Defra (2013); Cabinet Office Better Regulation Task Force, European Commission (2012, 2016):

a. *Develop regulations in collaboration with other countries* to minimise any possible competitive disadvantages in comparison with foreign companies not subject to the same regulations. Moreover, there is a need to share knowledge and experience (e.g., on best practice techniques) internationally through, for example, Impel and the Belfast group of UK regulators (see Chap. 1).
b. *Focus on outcomes, not prescribed technologies:* Regulations that prescribe specific technologies or are based on them can discourage innovation. Regulations should set out desired environmental outcomes and be flexible as to how best to achieve them.
c. *Have ambitious rather than lax objectives:* Regulations need to be strict enough to induce business to implement real innovation and technical changes to achieve them.

d. *Allow flexibility* for how to achieve these objectives and desired outcomes
e. The regulations need to be based on based on a *sound economic and financial appraisal* to ensure they are efficient and affordable and do not impose excessive costs on business so as to minimise any adverse economic impacts. This is in line with the requirements that regulators such as the EA and Natural Resources Wales report annually on the economic impact of their regulatory policies and practices.
f. *Integrate regulations in associated fields:* Porter and Linde (1995b: 111) stress the need to move away from single media and piecemeal approaches to integrated approaches, fundamentally geared to tackling all problems at a site which can lead to more innovative and efficient solutions. In addition, they stress the need to regulate as close as possible to the desired outcomes as possible, while flexibly encouraging all solutions—including upstream options. Avoiding pollution and addressing pressures at source is almost always less costly than end-of-pipe controls, remediation or clean up.
g. *Have well-defined phase in periods and deadlines:* Linking these to an industry's capital investment cycles can enable firms to integrate the environmental controls in investments for new plant and equipment which can facilitate development and adoption of new technologies that are better in economic, environmental and resource terms and less costly than adding end-of-pipe controls to existing plants.
h. *Make the regulatory process more stable and predictable* so that business can build in environmental considerations into their development, design and implementation of investments in new plants, equipment and technologies.
i. *Involve businesses in setting standards from an early stage:* There should be a more collaborative and less adversarial relationship between business and regulators. Business should focus their efforts, resource and attention on seeking innovative solutions to tackle environmental problems rather than contesting the regulations with expensive litigation. There should be greater positive use and attention to environmental managers and less resorting to (expensive) lawyers.
j. *Use market incentives*, such as pollution charges, deposit-refund schemes or tradeable permits, to provide continuing incentives for

business to develop and apply innovative solutions and go beyond just complying with any current standards (Wagner 2003). The impacts of tradeable permit schemes can significantly depend on the initial allocation of permits since the section " Impact of Environmental Regulation on Sustainable Economic Growth" highlights that granting (grandfather) rights to existing firms can adversely affect growth by preventing new firms (especially dynamic small firms) from entering the market.

k. *Minimise the time and resources involved in the regulatory process* so that it does not hinder or discourage innovation and investments.
l. *Enhance the technical capabilities of regulators* so that they are better able to understand the economics of the businesses they regulate and are better able to communicate with businesses regarding the best way to tackle the environmental problems in a well-reasoned and ordered way.
m. *Transparency:* The regulator needs to provide clear, credible and comprehensible information on the scale, nature and significance of the current and likely future environmental problems and pressures and their causes. This should be the essential rationale and basis for the environmental regulations. It can also provide business with the information needed to respond innovatively with long-term solutions rather than short-term compliance. Moreover, there needs to be clear transparent information on the measures implemented and their costs and benefits. Transparent exchange of such information internationally can provide good evidence of the extent to which foreign competitors are subject to similar requirements and costs so that complying UK firms will not be at a competitive disadvantage vis a vis foreign firms.
n. The environmental regulators need to have *strong links with the Industrial Strategy being developed by the Department for Business, Energy and Industrial Strategy and particularly government programmes to promote innovation* and application of best practice technical changes (e.g., through Innovate UK). Examples include:

- research in environmental technologies (e.g., The French Environment and Energy Management Agency (ADEME)'s 1.3bn € programme)
- guidance and information (e.g., the Austrian Ministry of Agriculture's Ökoprofit Programme)

- demonstration plants (e.g., Germany's Environmental Innovation Programme)
- verification and certification of new environmental technologies (e.g., the EA's ETV Programme and Germany's Blue Angel eco label)

A key question here is whether advice to businesses about environmental technologies is best provided by environmental regulators or by bodies promoting innovation and technical change by businesses or through industry association or by business advisers. Alternatively, by the close collaboration of all these parties—such as the waste minimisation clubs and resource efficiency clubs for specific sectors or regions (see Chap. 5 on the sector approach). Business need to determine which of these is best—since they have to pay for the service through either charges to recover regulator's costs or trade association contributions or direct payments to advisers. Public expenditure constraints could limit the extent to which any advice service by the environmental regulators could be funded by central government through Grant in Aid though it could fit well with the government's Industrial Strategy.

The EU Water Framework Directive (WFD) is now the main basis for water regulations in the UK and EU. It is well in line in the following ways with most of these principles:

a. *The WFD was developed in close collaboration with other EU countries*—at the UK instigation (see Section 2.6) and with considerable UK regulators' expert technical and economic help. The WFD is a valuable model for efficient integrated water management by other countries outside Europe who can benefit by going directly to integrated water management. This provides good potential exports for UK environmental services to aid these countries improve their water management.
b. *The WFD focuses on environmental outcomes* specified in terms of Good Ecological Status (GES) or Good Ecological Potential that is appropriate to particular types of water bodies in specific locations.

c. *The WFD aims to achieve ambitious GES objectives* which prompt businesses to implement real innovation and technical changes to achieve them.
d. *WFD gives member states flexibility* in drawing up their River Basin Management Plans (RBMPs) to aim to achieve efficiently these objectives and desired outcomes.
e. The WFD also explicitly *allows pragmatically for exemptions if achieving these objectives is not technically feasible or disproportionately expensive.* Moreover, the UK managed to secure inclusion of such consideration of costs and provision of exemptions in the implementation of the precautionary controls on Priority Hazardous Substances which could otherwise have been excessively expensive. The UK achieved this in close collaboration with other EU countries (e.g., the Netherlands) who were similarly concerned about these potentially high costs. This ensures that the WFD measures in the RBMPs are proportionate, affordable and do not impose undue costs on business. Accordingly, the Environment Agency (2015a: 17) has developed systematic economic appraisal processes for assessing the costs and benefits of options so as to select the most efficient and effective measures to improve the water body in question. These processes are essentially designed so that the EA's technical managers can apply them as an integral part of their determining the programmes of measures. These EA's technical experts are duly trained to apply these economic appraisal techniques.
f. *Integrate regulations in associated fields:* The WFD involves the integrated management of water bodies and catchments that addresses issues of water quality and water resources. It explicitly pulls together previously separate directives such as those concerning Bathing Waters, Shell fish waters and Natura 200. Moreover, the European Commission and member states are currently working to link implementation of WFD with the EU Marine Strategy Framework Directive and the EU Floods Directive. This is less costly than dealing with each of these issues separately and sequentially. It is also more efficient and can prompt more innovative responses in the programmes of measures

g. *Have well-defined phase in periods:* The WFD is currently being implemented in cycles of RBMPs that have to be reviewed and updated every six years. The EA has just developed the second cycle of RBMPs.
h. This helps *make the WFD regulatory process more stable and predictable.*
i. Article 14 of the WFD states that "Member States shall encourage the *active involvement of all interested parties* in the implementation of this Directive, in particular in the production, review and updating of the river basin management plans". The preparation of the 2015 update of the RBMP and recent periodic review of the water industry (PR14) embodied a more collaborative and less adversarial relationship between the OFWAT and environmental regulator (EA). Hopefully, water companies can then focus their efforts, resources and attention on seeking innovative solutions to tackle environmental problems rather than contesting the regulations with expensive litigation, which happened in earlier appeals on the previous technology-based Urban Waste Water Treatment (UWWT) Directive. As occurred, for example, in United Utilities' unsuccessful appeals against measures to control Combined Sewerage Overflows that were required under the UWWT Directive.
j. *Market incentives:* The WFD encourages the use of economic instruments. The environmental regulator and management body in Wales—Natural Resources Wales—have included trialling of a reverse auction scheme with businesses to address more efficiently the impact of high nutrient loading in a waterbody. Such reverse auction schemes involve asking polluters (e.g., farmers) to bid in terms of their costs for measures to control nutrients. The regulator then selects from the bids the most cost-effective suite of options to control nutrients. This is more efficient than just paying grants for the polluters to implement measures. Article 9 of the WFD specifically requires member states to report on the extent to which water prices cover the full financial, environmental and resource costs of water services. This encourages them to implement charges to cover the environmental and resource costs of water supply, abstraction and discharges. Moreover, and

potentially more important in practice, it requires that member states set out transparently their current subsidies in terms of the extent that water charges or prices do not cover the full financial costs of water services such as irrigation or abstraction of water. Such subsidies can distort competition. It puts at a competitive disadvantage UK businesses and farmers who rely more on rain and whose irrigated water supplies are not subsidised. World Wildlife Fund (WWF) in Spain carried out analyses which showed that strawberry farmers in Spain received significant water subsidies which exceeded their profits.

k. The WFD is often criticised for being a complex and time-consuming process because in reality achieving water management concerns many complex matters. But the *integrated regulation in the WFD takes less time and resources* than separate regulation of each water matter.
l. The EA had enhanced its teams' *technical capabilities* so that they are better able to derive plans for the best way of tackling water problems in their catchments.
m. *Transparency:* Potentially importantly, member states also will have to set out transparently the measures in their RBMPs and their costs and benefits and the extent to which they cover the full costs of water services (see above). UK environmental regulators have to justify the costs and benefits of measures to the regulated businesses paying for them. At the same time, they have to set out clearly how the costs of some measures are disproportionately expensive to justify to the European Commission and environmental non-governmental organisations (NGOs) as to why they have not implemented them and require exemptions from achieving good status. This alters the burden of proof to explain why the costs are too high, which should reduce tendencies to over-estimate costs.

The better-regulation initiatives described in other chapters substantially lowered environmental regulations' compliance costs for business and significantly improved their impacts on economic growth.

Foreman, in his chapter, shows how sector plans have developed to become a key part of how environmental regulators engage with industry. They aim to enhance understanding of sectors and their needs and constraints so as to be able to achieve more efficient and proportionate implementation of regulations. They can also enhance engagement at a high level in businesses that can prompt them to implement corporate strategies that achieve greater environmental improvements. The plans should highlight economic pressures on a sector which affects its ability to bear additional environmental control costs and might also highlight cases where the additional environmental control costs might be the "straw that breaks the camel's back" in terms of prompting plant closure and exit decisions. These can arise at critical times such as plant replacement for sectors where long-run profitability is low (e.g., dairy, steel and refining industries).

The following general economic intelligence can help environmental regulators to prepare their regulation strategies for key sectors such as refineries, electricity supply and chemicals. There are difficulties with providing detailed economic data for this since the available economic classification of sectors may not correspond precisely with the regulator's technical definitions of sectors in their strategies. Also, the economic data may not be readily available at a sufficiently disaggregated level to correspond with the regulator's specific technical definitions. So, this intelligence and analysis will need to be in terms of general trends rather than any detailed economic data on the sectors.

- Their economic context in terms of:
  - international competition and the UK sector's competitiveness with EU and non-EU countries (e.g., the USA, Middle East, Far East, Brazil, Russia, India and China (BRICs))
  - current situation and prospects for key factor inputs (e.g., energy/gas costs)
- the age of their plant and where the firms are in their normal asset replacement cycle and plans: Firms in the Netherlands have customarily carried out greater investment to replace their plants than their competitors in the UK. This raises questions as to the reasons for this, which are pertinent to not only environmental regulation but also more importantly to both businesses strategies and the government's Industrial Strategy.
- financial context regarding their access to capital and costs of capital (which can be significant for small firms—see May 2016)

- their market situation and the extent to which they operate in the global market and are more able to move plants and investments with changes in economic circumstances.

The sector plans can also highlight growth sectors (e.g., food and drink and agri food), where there is a need for more efficient technologies to enable these sectors to grow while safeguarding environmental conditions. Moreover, they can enhance knowledge about sectors where growth is directly stimulated by environmental regulation (see Sect. 1.3).

2.5 Uncertainties About Environmental Regulation

There are often calls to make environmental regulations more certain, stable and predictable. Thus, the Aldersgate Group (2012a) calls for "Credible, Consistent and Bankable Policy". But achieving "certainty" regarding future environmental regulations is in fact not feasible. Future environmental regulations depend on a balanced assessment of likely developments in the following factors about which there is inevitably considerable uncertainty:

I. The scale and nature of the environmental problems and pressures: Defra's National Ecosystems Assessment and the Natural Capital Committee have effectively highlighted the growing pressures on natural ecosystems which will require increasing environmental protection measures.
II. the public's concern and valuation of these problems
III. the costs and technical feasibility of control options, which depend on likely technological advances and innovations

Recent attempts to provide foresight and visionary "certainty" regarding future environmental regulations have been illusory and failed miserably because they did not adequately address all of these factors in determining the current regulations let alone future regulations. Box 2.1 gives an example of these shortcomings and their implications regarding the case of Feed-in Tariffs (FITs).

Box 2.1 Experience with feed-in tariffs

In April 2010, the Department for Energy and Climate Change (DECC) introduced feed-in tariffs (FITs) to support small-scale renewable energy installations with a capacity of less than five megawatts (MW). Under FIT, subsidies are paid for every kilowatt hour (kWh) of electricity generated and exported to the national power grid for 20 years. This long-time horizon was designed to give a clear long-term signal to encourage the development and application of renewable energy technologies, in which it succeeded.

The original policy and schedule of payments for FITs were based on standard economic appraisal in line with standard discount rates. However, it was not based on an adequate *financial* appraisal. In particular, it took insufficient notice of the actual costs of capital for the application of the technologies.

DECC (2012, p. 11) reports that the actual uptake of FIT schemes for small hydro-schemes (of < 15 kw) is lower than their predictions at the start of the FIT scheme—by 33% in numbers of schemes and 50% in terms of MW of energy generated. This is a fair reflection of the situation on the ground. This was due to the significant difficulties in securing funding for small hydro-schemes. Conversely, the uptake of solar PV far exceeded their forecasts because householders could fund the schemes at low capital costs (through mortgages). Consequently, overall uptake was much greater than predicted.

This meant that DECC was in danger of exhausting the available budgeted funds and therefore had to cut FIT rates which led to outcry by the renewable energy industry.

Morgenstern (2016: 10) states "Uncertainty is prevalent in everyday life, and it's no less prevalent in the regulatory world". Martin Bigg, in Chap. 1, states "In an uncertain world, one certainty is change". Business has always to deal with uncertainty—including on much more important and volatile factors such as the price of oil and exchange rates. Therefore, what is needed is "less uncertainty" and more stable and predictable environmental regulations—rather than trying to provide "certainty" about future regulations. This can best be achieved by the environmental regulators openly providing clear and credible information on the current situation and likely future developments regarding the scale and significance of environmental problems and pressures and public concerns about them—and hence the need for (stricter) environmental regulations now and in the future (see factors I and II above). Enhancing State of the Environment reporting could be a useful vehicle for this. This could give a good clear information and signal to business who are best placed to assess likely developments in technologies and the costs of control options, which will input into determining future regulations (see Factor III above)—in line with principles set out in Section 2.4.

2.6 Recent Political Perspectives on EU Environmental Regulation

As Martin Bigg shows in Chap. 1, the UK has had a major and increasing influencing in shaping European environmental regulations over the last 20 years. Thus, the UK developed, in 1991, integrated pollution control (IPC) for integrated regulation of discharges to air, water and solid wastes from major installations. Subsequently, the EU introduced the Integrated Pollution Prevention and Control (IPPC) Directive in 2008 (Office of Journal of the European Union 2008). Kestersen and Bigg discuss these directives in their chapters.

As a result, EU environmental policies have increasingly incorporated many of the features highlighted in Sect. 2.4 that help improve the impacts on economic growth. European environmental regulations are now to a much greater extent in the form of "framework" directives that aim to achieve specific environmental objectives and outcomes but give member states flexibility to determine how to achieve them efficiently.

Developments in European water policy are a particular notable example, which I focus on here. In the 1970s, there was a strong debate whether water regulations should be based on the "Uniform Emissions Standards" approach, advocated mainly by Germany on the continent or Environmental Quality Objectives (EQO) advocated by the UK. The former technology-based regulations involve uniform standards for each firm or plant in an industry regardless of their location. This held sway in the 1970s, in the form, for example, of the EU UWWT Directive, which set standards for discharges from waste water treatment plants and sewerage systems so as to make them fit for the end of the twentieth century and to reduce their significant pollution of water bodies. But the costs of these standards were very high. Moreover, they led water companies to move to large sewage treatment plants at the bottom of catchments, which took water away from the top of catchments and significantly increased water resource problems there.

In the late 1990s, the European Commission developed the EU WFD, which has the major positive features described in Sect. 2.4. It was implemented in 2000 (see Official Journal of the European Communities 2000). It superseded the earlier specific technology-based standards such as the UWWT, though member states still had first to achieve compliance with these existing directives and then go on to implement the WFD to improve their water management more effectively and efficiently. It fundamentally embodies and enshrines the principles of the UK's EQO approach. The UK played a big role in the development of the WFD, which at that time was called the "English Directive" by many on the continent. For example, Pamela Taylor (Water UK) in her evidence to the House of Commons Select Committee on Food and Rural Affairs (2003: para 324) stated that "traditionally we have done a lot of our work based on catchments in particularly England and Wales, so we felt that we had a head start. And it is interesting, I am sure you may know, that in Europe this Directive is often referred to as the British Directive, or the English Directive". Similarly, Helen Rimmer (2003) stated that "The government has never had a better chance to address such problems as it finalises details for implementation of the European WFD. Known in Europe as the 'English' directive, because it was developed largely by British scientists, it has been hailed as the most ambitious piece of European environmental legislation ever". Similarly, Jacob Tomkins (2016) of Water Wise stated that "the Water Framework Directive, the largest piece of environmental legislation ever, which aimed to ensure good quality water across Europe and was adopted by countries worldwide (so even if we have left Europe and are drifting in the Atlantic when you read this, it is still relevant). This policy was led by the UK, and its nickname was 'the English Directive'".

It was ironic that, at first, in preparing the first RBMP cycle in 2009, Defra and the EA had great difficulty in implementing its "English" Directive in England and Wales because they implemented it nationally rather than at the River Basin District or catchment level as the WFD requires. This was because of limited capability and capacity at catchment level, and the high costs of the outstanding requirements for the existing directives (e.g., UWWT and Habitat) meant that there was little room left for affordable WFD measures.

The Angling Trust and WWF took Defra to judicial review because they claimed that the first RBMPs did not fully comply with the requirements of the WFD (see Angling Trust (2010)). However, this seriously diverted scarce resources at Defra and the EA from important work to prepare measures to implement the WFD. A less adversarial approach would have been more productive and environmentally beneficial. Moreover, it would have been better if they had focused their attention more on working internationally to highlight the much greater inadequacies in the implementation of the WFD in other EU countries—as they did well in highlighting the significant distortions caused by subsidised water in Spanish strawberries (see Sect. 2.4).

In 2013, Defra launched the Catchment Based Approach which formed the basis of their preparing the second cycle of RBMPs in line with the WFD.

The WFD is criticised for being too complicated. However, effective and efficient water management inevitably covers many complex matters. In addressing them, the WFD does all the right things in the right way and that is inevitably complex. This means that it takes time to develop appropriate processes for preparing RBMPs to implement the WFD. It will not be until the third cycle of RBMPs that Defra and the EA can get it right.

When the WFD was being negotiated in 2000, people thought 2027 seemed a long time away and rivers were improving, so, GES by 2027 seemed realistic. Now, however, people in all member states realise that GES is not at all easy to achieve everywhere in these timescales. Therefore, it will be essential that WFD's initial 2027 deadline is extended so that less stringent objectives can be permitted well-beyond 2027 since this deadline would either not be technically feasible or would be excessively costly. This extension would enable the EA to continue to build on its current good work to achieve efficiently and effectively significant improvements to the water environment.

Issues Regarding Brexit

Following the referendum decision on 23 June, the key question now is what sort of Brexit is best for the UK and the EU and accordingly what degrees of "hard" or "soft" Brexit that the UK Government goes for and can negotiate during the two-year period after Article 50 was invoked at the end of March 2017. Key issues surrounding this question and debates include:

a. Whether the UK can retain "friction free" access to the EU single market—that is, access that current EU member states have without tariffs or customer's barriers. The EU single market is the largest market in the world. It currently accounts for 44% of UK trade. Exports to the EU currently account for 12% of UK GDP. Loss of such free access to the single market could result in significant losses to sectors currently exporting to the EU. The Economist (2016b: 28) reports estimates by Oliver Wyman that losing such free access to the single market could cost 35,000 jobs in Finance. Moreover, business investment could decline due to uncertainties and lost confidence arising from Brexit and reductions in foreign direct investments by firms that have traditionally been located in the UK to be able to trade easily in EU's single market.
b. Control on immigration from the EU was a key issue behind the referendum result and is likely to cause an impasse in the negotiations on Brexit. EU member states are demanding agreement to freedom of movement of people as a key core freedom and a sine qua non-requirement for free access to the single market, while the UK demands controls on EU immigration. It appears that neither side is prepared to derive a practical pragmatic compromise. It is ironic that, having campaigned during the referendum for reducing the burden of EU red tape, hard Brexit proposals would impose bureaucratic immigration controls which would have much more adverse economic impacts on the ability of businesses to grow dynamically than any economic impacts of environmental regulations, which Sect. 2.3 showed to be not significant.

c. What regulations will the UK impose? Martin Bigg's introduction in Chap. 1 provides an overview of the issues and options. This chapter then examines this subject in more depth focusing on economic aspects first in the context of overall environmental regulation and then in the specific context of the latest EU water regulation—the EU WFD.

Environmental Regulations and Brexit

During the referendum, there was little reference specifically to environmental regulations, which means that that there is not really a mandate for any knee-jerk dismantling of environmental regulations post-Brexit. Only reference I can recall was a bogus claim that the WFD gives preference to the environment over people and this led to flooding. The WFD and EU Floods Directive give full consideration to both people and the environment. Reduced flood risks to people and businesses account for the majority of the benefits that the EA takes into account in determining flood risk management measures.

On Brexit, in order to reduce uncertainties for both regulators and business, an essential immediate legal task is to convert the current body of EU law into UK legislation with changes to the law so that it continues to work as it currently does (e.g., where regulation currently relies on EU systems/institutions/quotas, etc.). There will not be any significant immediate changes to environmental regulation, including the provisions of the WFD. The exceptions are replacing the Common Agricultural and Fisheries Policies, which will require legislation in 2017.

There are criticisms that this (misleadingly named) "Great Repeal Bill" fundamentally goes against the objectives of and reasons for Brexit to break free from EU regulations (e.g., Shapps 2016). Consequently, Shapps calls for a five-year sunset clause in this Bill to allow MPs to scrutinise former EU laws, removing job-destroying clauses. However, this would create considerable uncertainties for business concerning what will be the regulations after this sunset period. It would also create considerable legislative burden and logjam. Even extending the sunset clause to ten years would still divert parliament from developing and improving other more important legislative changes. We need a more focused approach.

Depending on the UK's new relationship with the EU arising from the Article 50 negotiations and as part of transitioning the exit from the EU, Defra will need to review environmental regulation in a measured and focused way in the context of Defra's forthcoming 25 Year Environment Plan, which should be published by the end of 2017.

We must retain as much stability and predictability regarding environmental regulations so as not to create unnecessary uncertainty which would have adverse economic impacts on business. Moreover, it needs to be focused and based on sound overall evidence—not dogma and anecdotal misinformation. We must not throw the baby out with the bath water.

Consequently, this brief review should cover the following matters:

a. Identify those EU Directives and regulations from which we have exemptions and hence would not need to be transposed into UK law. So, take these out of the debate. These matters should be covered by any current UK laws and regulations that we apply instead.
b. Focus on specific concerns about the remaining EU laws currently applicable here.
c. Determine whether these concerns could best be handled by amending the targets and judicious efficient implementation in line with our current better regulation principles (rather than legal changes as such). I believe that this could address many of the concerns in the area of EU environmental regulations which are largely "framework" directives and give considerable flexibility in how the member states implement them (see Sect. 2.6 earlier) especially regarding the WFD.
d. Highlight actual hard evidence of those cases where EU regulations impose excessive costs and destroy UK jobs. This would also usefully show whether the concerns during the referendum about bureaucratic EU laws are based on misinformation or actual facts in the light of this proper scrutiny of the evidence.
e. Accordingly, focus on these outstanding EU laws that require legal amendment. Then, have a measured 5–10-year plan for making these

amendments in a sensible staged way, focusing first on the most important ones.

f. Set out the scale, nature and significance of the costs, especially costs to business of these regulations requiring amendment. Brexit means that there will be increasing attention concerning the economic impacts of regulations for political reasons—given demands to reduce the burden of EU red tape—and economic reasons. The UK is likely to face difficult economic challenges following Brexit. Moreover, there will be increasing attention concerning impacts on international competitiveness, especially concerning countries outside the EU and any potential for UK firms to relocate investments to non-EU countries with lower environmental standards.
g. Set out the scale, nature and significance of the environmental problems that the amended regulations tackle and the benefits of the regulations. Brexit means that the environmental authorities (and also the environmental pressure groups) will have to move from focusing on compliance with EU Directives to seeking justifiable environmental improvements that are worthwhile for the UK (i.e., that their benefits exceed their costs). This is all to the good and consistent with the best practice principles outlined in Sects. 2.4 and 2.7 and with which most current environmental regulations are well in line.

The assessment in (g) above should help to show that environmental regulations are addressing major environmental problems and also that these problems are likely to be even more significant in other non-European countries with currently lower environmental standards, who will therefore need to implement tighter environmental regulations in future. This will reduce any tendency for business to relocate to such current pollution havens (see the section "Impacts on International Competitiveness and Pollution Havens").

Where an environmental regulation does not allow (adequately) for the costs of complying and where their targets are not currently feasible or too costly to achieve and not sufficiently beneficial and worthwhile, then they should be treated as "aim to achieve" objectives, requiring proportionate measures to be implemented over reasonable time periods.

Any such review should assess the UK's implementation of EU directives in terms of the better regulation principles outlined in Sects. 2.4 and

2.7. Items (f) and (g) would need to be covered in the Impact Assessments that Defra would need to provide as part of the regulatory reviews and to underpin any regulatory and legislative changes. Defra needs to pay particular attention to any reviews that the European Commission is itself carrying out regarding how their own regulations need to be amended to better fit their own guidelines.

It will be essential that, as far as possible, the UK continues to play its traditional role of positively influencing and shaping environmental regulations that it has so successfully carried out over the last two decades. In particular, the UK authorities need to ensure that the EU targets and environmental regulations are based on a sound economic and financial appraisal and that the targets and deadlines are feasible and worthwhile in achieving important benefits and not entailing disproportionate costs.

There will be strong pressure from the EU to make any continued current frictional-free access to the EU's single market conditional upon UK complying with EU regulations—as in the European Economic Area model (see IIEP 2016a, b)—to retain an even-playing field across Europe. Therefore, it will be important that the UK continues to influence positively EU environmental regulations during the Brexit negotiations and thereafter. We must retain those environmental regulations and processes that are well in line with these principles and are essentially "British" in basis.

The Water Framework Directive and Brexit

Brexiteers gave rhetorical arguments for Brexit during the referendum debates and now for a "hard" Brexit, which superficially appear convincing. In the context of the water sector, Stephen Topping (2016) sets these out eloquently as including:

A. Brexit could lower the water sector energy costs if it did not have to implement costly EU targets for increased renewable energy.
B. Brexit could increase the sector's ability to hire skilled labour from non-EU countries.
C. Brexit gives the UK the opportunity and flexibility to decide what EU regulations are best for the UK and thus what it wants to adopt in what form and what it does not.

D. Brexit provides opportunities for UK companies to export water infrastructure and engineering services to non-EU countries, which is a major world market.

I examine each of these arguments in turn.

A. lower costs:

Horton (2016) acknowledges Topping's point A above but counters that a hard Brexit could in fact lead to higher energy costs for the water sector since it might mean that the energy sector would no longer be subject to EU competition laws, which, in the absence of equivalent UK measures, would reduce competitive pressures to contain or lower energy prices in the UK. Also, the UK is a net energy importer. A hard Brexit could weaken the cross-border energy market and make cross-border arrangements (e.g., the cross-channel interconnector) more difficult. Moreover, falls in the pound following Brexit would raise the cost of imported energy.

In addition, Horton (2016) argues that Brexit will lead to lower investment in the UK water sector due to the loss of the major investments that the European Investment Bank (EIB) has provided for the UK water sector. In the latest periodic review of the water industry in 2014, prior to Brexit, the water industry's cost of capital was at an all-time low of about 3.7% in real terms (Ofwat 2014: 10). The water industry's cost of capital is likely to rise following Brexit due to this loss of EIB finance and increased uncertainties arising from Brexit. Also, if the fall in sterling following Brexit leads to a rise in inflation, then the Bank of England might have to raise interest rates to contain inflation to within its 2% target. This could further increase the pressure on the sector's cost of capital. The water industry's capital costs account for about half of its total costs, so, this could increase water prices for the consumers.

B. hiring skilled labour:

Horton (2016) stresses that the UK water sector (along with other business sectors in the UK) currently is severely short of skilled labour.

Excell (2017) states that the engineering sector will require 265,000 engineers p.a. in the UK over the next 10 years. At present, this gap is largely filled by EU nationals. Therefore, strong immigration controls from the EU would damage this sector in terms of loss of access to this skilled workforce and/or bureaucratic regulatory costs and delays as they have to seek work permits for these staff to fill the posts for which there are currently insufficient UK employees. Moreover, it is not evident that the UK immigration policy post-Brexit would lead to increased ability to hire skilled labour from outside the EU to fill the shortfall. Furthermore, it would appear that the water sector (as with other sectors) would still incur the regulatory costs and uncertainties of having to overcome bureaucratic hurdles to obtain work permits for these staff. Brexit will necessitate great concerted action to improve training of the UK labour force to fill this skills shortage—though this is needed anyway regardless of Brexit. But it will take time for such action to lead to significant increases in skilled workers. Hence, a key question for the government and Brexiteers is how to handle the significant skills shortages that UK businesses will face in the interim.

C. allow the UK to choose flexibly the regulations that are best for the UK:

Fisher (2016) agrees with Topping here. But this should not mean dismantling the WFD, about which we need to retain with its sensible "English" processes. Recent RBMPs will achieve important environmental benefits (with a present value of about 22.5 bn) that exceed their costs (with present value of 17.5 bn) (Environment Agency (2015b).

However, it will not be feasible to achieve GES by 2027. Consequently, this current EU deadline should be extended and we need to treat GES as an objective which various subsequent rounds of RBMPs will aim to achieve. This is also the view of others in EU member states. So, it should happen anyway—regardless of Brexit. Defra need to link with, inform and influence any such reviews of the WFD, its GES target and deadlines that the European Commission is itself carrying out. We should also promote efficient ways such as abstraction trading to enable compliance with the WFD's "no deterioration" objective without restricting development.

D. export opportunities for the water sector in non-EU countries:

Fisher (2016) acknowledges that the market for water supply and waste water treatment infrastructure in non-EU countries is big, although it can be limited due to existing low-water prices there and affordability constraints that limit the ability to raise water prices. The UK water companies have historically not exploited this market, which is now dominated by incumbent French and German water companies who will strive hard to keep out any new competitors from the UK.

There is greater potential for UK companies in providing integrated services to help non-EU countries develop and apply new, more efficient water management policies, regulations and systems. These countries should go directly to integrated water management based around the "English" WFD and the Floods Directive (see Sections 2.4 and 2.6). The UK water companies and consulting services ought to have a strong comparative advantage in exploiting this market. However, if hard Brexit means abandonment of the WFD and exclusion of UK companies from engaging with European partners on its development and implementation, then these UK companies would rapidly lose this essential advantage and potential exports.

To sum up, this review of the evidence suggests that a hard Brexit would lead to a loss of essential skilled labour and increase labour, energy and capital costs for the UK water sector and lead to a loss in future exports to both EU and non-EU countries rather than the potential gain advocated by Brexiteers. Brexit should lead to desired changes in how the WFD will be implemented. But these changes should happen anyway. The UK needs to retain the WFD and continue to influence how the EU should sensibly refine and implement this "English" Directive.

2.7 Summary, Outlook up to 2025 and Conclusions

The available evidence on the economic impacts of environmental regulations is somewhat contested and not perfectly clear-cut. Nevertheless, the following clear conclusions can be drawn from the available evidence:

1. The weight of evidence suggests that there are no *significant* economic impacts of environmental regulations. Environmental regulations' gross costs to all businesses account for just 0.16% of the turnover of all businesses and only 0.2% of the turnover of manufacturing businesses. The net costs to business of environmental regulations are generally about one-third lower when account is taken of the direct financial benefits of environmental regulations to business.
2. Better regulation initiatives can significantly improve the impacts on economic growth and positive impacts on the UK Environmental Goods and Services Sector (UK EGSS), which is an important growing UK industry catering for a market that is increasing significantly especially in Europe and worldwide (see Sect. 2.2.3).

Increasing environmental pressures and scarcity of important ecosystems services and rising public concerns about environmental matters (see Defra 2011, 2014; Natural Capital Committee 2014) are likely to lead to the need for stricter environmental regulations in future. At the same time, the UK will face increasingly difficult economic challenges. There will be increasing concern about the economic impacts of environmental regulation.

Nevertheless, greater environmental protection can still be achieved without significant adverse economic impacts. To that end, Sect. 2.4 shows that the design and implementation of the environmental regulations need to be based on the following 16 best practice principles to improve their economic impacts. These are similar to those in other reviews and guidance (e.g., Defra (2013); Cabinet Office Better Regulation Task Force, European Commission 2012, 2016):

A. Collaborate with other countries, especially in Europe, and share best practice.
B. Focus on achieving environmental outcomes and not any prescribed technologies.
C. Have ambitious rather than lax objectives to provide an ongoing incentive to promote innovations to achieve the strict regulations and their desired outcomes.

D. Allow flexibility on how to achieve these objectives and desired outcomes.
E. At the same time, take due considerations of the costs and feasibility of the regulations and allow explicitly for exemptions where the costs can be shown to be disproportionately expensive. In this way, ensure that the regulations do not impose an undue cost burden on businesses.
F. Base the regulations and their implementation on a sound economic and financial appraisal of their full costs and benefits with particular attention to impacts on small firms. In the UK Government and European terminology, these appraisals are called "Impact Assessments". Defra and the EA need to evaluate performance of regulations and carry out empirical analyses of actual ex-post costs versus ex ante estimates for regulations which the Impact Assessment showed could have major costs for businesses. They need to learn from these evaluations and incorporate their findings in appraisals of future regulations.
G. Be proportionate. Ensure that the measures and their costs are proportionate to the significance of the problem and worthwhile in terms of the benefits they achieve.
H. Integrate regulations in associated fields to promote more innovative and efficient solutions that address various related environmental problems in one go and provide a one-stop shop to interact with business on the implementation of the regulations. Integrated solutions and avoiding pollution and addressing pressures at source is almost always less costly than end-of-pipe controls, remediation or clean up.
I. Have well-defined phase in periods. Linking these to an industry's capital investment cycles can enable businesses to integrate the environmental controls in investments for new plant and equipment which can prompt development and adoption of new technologies that are better in economic, environmental and resource terms and substantially less costly than adding end-of-pipe controls to existing plants.
J. Make the regulatory process more stable and predictable so that business can build in consideration of environmental controls into their

development, design and implementation of investments in new plants, equipment and technologies. This is best achieved by the regulators providing clear, credible and comprehensible information on the current situation and likely future positions regarding the scale and nature of environmental problems and pressures and public concerns about them—and hence likely developments in the need for (stricter) environmental regulations. This could give a good clear signal to business who will be best placed to assess likely developments in technologies and the costs of control options which the regulators will need to take into account to determine future regulations in a sound balanced way in line with these principles.

K. Involve businesses, environmental NGOs and stakeholders at an early stage in drawing up the regulations and setting the targets. There needs to be a more collaborative and less adversarial relationship between these various bodies.
L. Use market incentives to provide continuing incentives for business to develop and apply innovative solutions.
M. Minimise the time and resources consumed in the regulatory process so that it does not delay and hinder or discourage innovation and investments.
N. Enhance the technical and economic capabilities of regulators so that they are better able to understand the economics of the businesses they regulate and are better able to communicate with businesses regarding the best way of tackling the environmental problems in a well-ordered way.
O. Provide clear guidance for business regarding the regulations and how firms can best achieve the targets.
P. Transparency regarding the measures and their costs and benefits not only in this country but also in competing countries, especially in Europe.

Notes

1. This is all Defra regulations, which are a good proxy for environmental regulations but they include farming regulations and exclude carbon schemes.
2. House of Commons Treasury Select Committee (2011) found that the costs of capital of PFI schemes were also high—about 8% in real terms. There is little or no evidence of the PFI waste incinerator schemes yielding efficiency savings that could offset these high capital costs so that these PFI schemes were considered excessively expensive.

References

Aldersgate Group. 2011. *Dealing with Deficits: Best Value Regulation to Reduce Our Environmental and Financial Debts*. Aldersgate Group, June.

———. 2012a. *Greening the Economy: A Strategy for Growth, Jobs and Success*. Aldersgate Group, June.

———. 2012b. *Skills for a New Economy: A Paradigm Shift in Education and Learning to Ensure Future Economic Success*. October.

Ambec, Stefan, Mark A. Cohen, Stewart Elgie, and Paul Lanoie. The Porter Hypothesis at 20: Can Environmental Regulation Enhance Innovation and Competitiveness? *SSRN Electronic Journal*.

Angling Trust. 2010. *Government in the Dock over Illegal River Basin Plans*. March 26. http://www.anglingtrust.net/news.asp?section=29&itemid=524

Angus, A., Booth, C., Armstrong, G., & Pollard, S. J. T. (2013). Better evidence for regulatory reform: Rapid evidence appraisals. Contract report ERG117 prepared for the Department for Environment, Food and Rural Affairs (Defra) by the Centre for Environmental Risks and Futures, Cranfield University, Cranfield, Bedfordshire, UK, pp. 136 with appendices.

Cabinet Office Better Regulation Task Force. 2012. *Principles of Good Regulation*. UK Government Web Archive.

Department for Energy and Climate Change (DECC). 2012. *Feed-in Tariffs Scheme: Consultation on Comprehensive Review Phase 2B: Tariffs for Non-PV Technologies and Scheme Administration Issues*.

Department for Environment Food and Rural Affairs (Defra). 2011. UK National Ecosystem Assessment: Understanding Nature's Value to Society. 2 June 2011.

Defra 2013. *Smarter Environmental Regulation Review: Phase 1 Report: Guidance and Information Obligations.* May 16.

———. 2015. *Emerging Findings from Defra's Regulation Assessment, First Update Covering 2012.*

Defra. 2014. *UK National Ecosystem Assessment: Follow-on.* http://uknea.unep-wcmc.org/Resources/tabid/82/Default.aspx

Department for Business, Industry and Skills. 2016. *Template for Post Implementation Review.* http://uknea.unep-wcmc.org/Resources/tabid/82/Default.aspx

Economist. 2016a. *The Economist,* May 28, p. 24.

———. 2016b. *The Economist,* October 8, pp. 27–29.

Ederington, J., A. Levinson, and J. Minier. 2005. Footloose and Pollution-Free. *Review of Economics and Statistics* 87: 92–99.

Environment Agency. 2015a. *A Consultation on the Draft Update to the River Basin Management Plan Part 3: Economic Analysis—Extended Report.*

———. 2015b. *Update to the River Basin Management Plans for England's Water Environment.* https://www.gov.uk/government/publications/river-basin-management-plans-impact-assessment

European Commission. 2012. *Europe Can Do Better: Best Practice for Reducing Administrative Burdens.* European Commission, High Level Group of Independent Stakeholders on Administrative Burdens.

———. 2015. *Environmental Expenditures in EU Industries Time Series Data for the Costs of Environmental Legislation for Selected Industries Over Time. Final Report by TME.* Institute for Applied Environmental Economics, November. http://ec.europa.eu/environment/enveco/economics_policy/pdf/Costs%20of%20cnvironmental%20legislation.pdf

———. 2016. *Interinstitutional Agreement Between the European Parliament, the Council of the European Union and the European Commission on Better Law-Making.* s.l.: s.n.

Eurostat. 2016. *Environmental Goods and Services Sector.* http://ec.europa.eu/eurostat/statistics-explained/index.php/Environmental_goods_and_services_sector

Fisher, J.C.D. 2008. *Challenges for Applying Cost-Benefit Analysis and Valuation of Environmental Benefits to Aid Environmental Decision-Making in Practice.*

———. 2016. WFD and Brexit. Water Economics Forum Conference, organised by WRc. September.

HM Government. 2012. *Enabling the Transition to a Green Economy: Government and Business Working Together.*

HM Government. 2017. Building Our Industrial Strategy: Green Paper. January 2017. https://www.gov.uk/government/consultations/building-ourindustrial-strategy

HM Treasury. 2015. *Fixing the Foundations: Creating a More Prosperous Nation* [Cm 9098]. https://www.gov.uk/government/uploads/system/uploads/attachment_data/file/443897/Productivity_Plan_print.pdf

Horton, B. 2016. *The Potential Negative Impacts of Brexit for Water in the UK.* Water Economics Forum Conference,organised by WRc. September.

House of Commons Environment, Food and Rural Affairs Select Committee. 2003. *Fourth Report. Examination of Witnesses Questions 320–339.* https://www.gov.uk/government/uploads/system/uploads/attachment_data/file/443897/Productivity_Plan_print.pdf

House of Commons Treasury Select Committee. 2011. *Seventeenth Report: Private Finance Initiative.* http://www.parliament.uk/business/committees/committees-a-z/commons-select/treasury-committee/news/pfi-report/

Institute for European Environmental Policy (IEEP). 2016a. *The Potential Policy and Environmental Consequences for the UK of a Departure from the European Union.* http://www.ieep.eu/assets/2016/IEEP_2016_Brexit_-_Implications_for_UK_Environmental_Policy_and_Regulations.pdf

———. 2016b. *Brexit—The Implications for UK Environmental Policy and Regulation. A Special Independent Report Commissioned by the All-Party Parliamentary Environment Group.* http://www.ieep.eu/assets/2016/IEEP_2016_Brexit_-_Implications_for_UK_Environmental_Policy_and_Regulations.pdf

Mani, M., and D. Wheeler. 1998. In Search of Pollution Havens? Dirty Industry in the World Economy, 1960–1995. *Journal of Environment and Development* 7: 215–247.

May, T. 2016. *CBI Annual Conference 2016: Prime Minister's Speech.* https://www.gov.uk/government/speeches/cbi-annual-conference-2016-prime-ministers-speech

Morgenstern, R. 2016. *The Real Costs and Benefits of Federal Regulations: Interview with Richard Morgenstern.* Resources, No 191, Winter.

Natural Capital Committee. 2014. *The State of Natural Capital. Second Report to the Economic Affairs Committee.* p. 56. https://www.gov.uk/government/speeches/cbi-annual-conference-2016-prime-ministers-speech

Natural Environment Research Council (NERC). 2012. *Most Wanted II: Post Graduate and Professional Skills Needs in the Environmental Sector.* http://www.nerc.ac.uk/skills/postgrad/policy/skillsreview/

Official Journal of the European Communities. 2000. *Directive 2000/60/EC of the European Parliament and of the Council, of 23 October 2000, Establishing a Framework for Community Action in the Field of Water Policy.*
Official Journal of the European Union. 2008. *DIRECTIVE 2008/1/EC of the European Parliament and the Council of 15 January 2008, Concerning Integrated Pollution Prevention and Control.*
Ofwat. 2014. Setting Price Controls for 2015–20: Overview. December 2014.
ONS. 2015. *UK Environmental Accounts: 2015.*
Pepsico. 2010. *Water Stewardship: Good for Business. Good for Society.*
Porter, M., and C. van der Linde. 1995a. Green and Competitive. *Harvard Business Review*, September–October, 120–134.
———. 1995b. Toward a New Conception of the Environment-Competitiveness Relationship. *Journal of Economic Perspective* 9 (4): 97–118.
Rimmer, H. 2003. Cleaning Up Our Act; The Guardian 25 June 2003. https://www.theguardian.com/society/2003/jun/25/guardiansocietysupplement.water
Sato, M. 2014. *The Impacts of Environmental Regulations on Competitiveness.* London: London School of Economics and Political Sciences.
Shapps, G. 2016. Put Wealth and Jobs First. *The Times*, October 24.
Tomkins, Jacob. 2016. Water and the Big Bangers, *The EMA Magazine*, pp. 10–11.
Topping, S. 2016. *Brexit—The Opportunities for the Water Sector.* Water Economics Forum Conference, organised by WRc. September.
Wagner, M. 2003. *The Porter Hypothesis Revisited: A Literature Review of Theoretical Models and Empirical Tests.* Lüneburg: Centre for Sustainability Management.

Dr Jonathan Fisher is a freelance environmental economist with more than 35 years' experience of delivering practical applied economic analyses of environmental matters. Until 2014, he was Economics Manager at the Environment Agency responsible for economic analysis on water and flood risk management and implementing the EU Water Framework Directive. Previously, he worked as an economic adviser at HM Treasury and before that at the Department of the Environment, where he worked in the think tank that developed Integrated Pollution Control. He has a PhD in environmental economics and a degree in economics and accounting.

3

Better Regulation Initiatives

Chris Booth

3.1 Introduction

When introducing a subject, one often begins by trying to define it. This might be sensible with a scientific term, like gravity or metabolism, but it would be a mistake to try to do so with a concept like better regulation. There is no set of features that exclusively defines a set of laws or their implementation as "better regulation". The different laws and regulations and their practical implementation, enforcement, and so on that are placed in this category enjoy many resemblances of one to the other, but there is not a universal set of traits that is common to all of them.

Furthermore, once you start to try to define better regulation, one naturally comes up against the problem of defining "better than what?" or of justifying that "better" regulation is preferable to plain "same old" regulation or "worse" regulation, which hardly needs saying. So, perhaps an easier task is to describe it in terms of the *need* for better regulation. A succinct UK example is shown in Box 3.1 and a European Union example in Box 3.2.

C. Booth (✉)
Regulatory Specialist, Bristol, UK

J. Foreman (ed.), *Developments in Environmental Regulation*, Palgrave Studies in Environmental Policy and Regulation, DOI 10.1007/978-3-319-61937-8_3

Box 3.1 The Need for Better Regulation

Regulation has many purposes, including protecting consumers, employees and the environment, promoting competition and supporting economic growth. Regulation can benefit both businesses and consumers through, for example, building consumer confidence in the products and services they buy. However, businesses incur costs in complying with regulations, which can act as a barrier to competition and reduce productivity (House of Commons Committee of Public Accounts, 2016).

Box 3.2 Better Regulation Why and How

The better regulation agenda is about designing and evaluating policies and laws transparently, with evidence, and backed up by the views of citizens and stakeholders. It covers all policy areas and aims for targeted regulation that goes no further than required, in order to achieve objectives and bring benefits at minimum cost. (European Commission 2016a)

So instead of spending time and effort on definitions of better regulation, this chapter explores the types and characteristics of better regulation initiatives that have actually been implemented. For the purposes of this chapter, like the rest of the book, the emphasis will be on regulation of *business*[1] rather than of individuals or households, and the focus is on *environmental* regulation of business. Having said that, this chapter starts with *wide-ranging* initiatives of better regulation of business before moving on to those initiatives related exclusively to *environmental* regulation. It covers the significant developments in better regulation in the European Union (EU) and provides more detail on specific UK initiatives.

It is perhaps worth clarifying, before going any further, that many better regulation initiatives discussed below have not actually been labelled as *better regulation* but have come under various similar headings such as those in Box 3.3. No doubt the individuals and organisations who chose such terms rather than using *better regulation* have reasons for doing so, and maybe they might argue there is a difference in their meanings. However, for the purposes of this chapter, the author has included initiatives under such headings where they have attempted to change regulation for the better.[2]

Box 3.3 Some other terms often used in place of "better regulation"

Deregulation

Modern Regulation

Smart Regulation

Regulatory Reform

Austerity Regulation

Inspection Reform

Good Regulation

Future Regulation

3.2 Types of Better Regulation Initiatives

Most wider better regulation initiatives have generally started from a political perspective based on the assumption that although regulation of business has its place it should not be unduly burdensome so as to add costs to business, prevent growth, or make business start-ups unreasonably difficult. Often this is accompanied by a recognition that small and medium-sized businesses are proportionally more affected by such "regulatory burdens" than larger businesses.

These initiatives have included one or more of the following aspects:

- New legislation: scrutiny of any new proposals to determine, for example: the regulatory impact of the legislation; whether or not it is needed; whether or not policy goals could be achieved by other means. For example, the Regulatory Policy Committee was set up at the end of 2009, to provide external scrutiny of the policy-making process in England and Wales.
- Existing individual pieces of legislation. To review and remove if they are unduly burdensome to business and/or ineffective, etc. An example was the Red Tape Challenge (RTC)—a comprehensive government

review of 21,000 regulations on different themes. As a result, certain specific items of legislation were revoked such as the Site Waste Management Plans Regulations (HM Government 2008) which made it compulsory for construction projects over £300,000 to produce site waste management plans. This was revoked in 2013 (HM Government 2013).

- Overall *categories* (rather than individual pieces) of legislation applying to business in a particular category such as health and safety law, employment law or environmental law. The purpose of this can be to establish how legislation as a whole is operating in any given area. Reviews might cover for example: to what extent the legislation appears to be a coherent integrated whole; whether it is consistent in its terminology, philosophies, descriptions of offences and penalties, reporting requirements, and so forth; and whether it can be easily interpreted by users. The purpose here being, where appropriate, to simplify, streamline, integrate, and so on. An example is the smarter environmental legislation project in the Department for the Environment, Food and Rural Affairs (Defra) (Department for the Environment, Food and Rural Affairs 2014). This is part of the Smarter Environmental Regulation Review in Defra (Department for the Environment Food and Rural Affairs (Defra) 2012). This is described more fully in chapter 4.
- Implementation of Legislation. Better regulation initiatives in this area have looked at the application of regulation in practice. Aspects covered include reporting, guidance, enforcement mechanisms, risk-based allocation of inspection resources and alternative approaches to inspection (inspection reform). An example is the integrated Environmental Permitting Regime, developed and implemented in stages by Defra and the Environment Agency (EA) from 2005 to 2013. The regime covers facilities previously regulated under the Pollution Prevention and Control Regulations 2000, and Waste Management Licensing and exemptions schemes, some parts of the Water Resources Act 1991, the Radioactive Substances Act 1993, and the Groundwater Regulations 2009 (Department for Food, Environment and Rural Affairs 2013). Another example is the work done by the European Union Network for the Implementation and Enforcement of

Environmental Law (IMPEL) on complementary approaches to inspections described below.
- Evaluation: Either "ex-ante" during design stage before implementation and or "ex-post" after implementation. For example, the EU undertakes evaluations and "fitness checks" on EU legislation (European Union 2015).

Such better regulation initiatives are not necessarily solely confined to environmental regulation, but cover wide-ranging aspects of what has been called *social regulation* such as health and safety, finance, employment, gaming, and so on. The common element of social regulation is that it seeks to intervene in the affairs of business so that business operations do not adversely affect aspects of wider society such as health and safety, employees' rights, vulnerable individuals, and of course the environment. Examples of *wide-ranging* better regulation initiatives that affect such a broad range of social regulation are discussed below particularly where they have had a bearing on *environmental* regulation. Later sections describe some specifically *environmental* better regulation initiatives.

3.3 Wide-Ranging Better Regulation Initiatives

Better Regulation in the UK 1997–2010

Following the 1997 General Election of "New Labour", and over the next ten years or so, the UK Government established a number of bodies charged with better regulation duties and initiatives to see these through. The scope of these was wider than (but included) *environmental* regulation. The most significant of these bodies and initiatives were:

1. *The Better Regulation Task Force* which quickly devised the following five principles of "good" regulation:

    - Proportionality
    - Accountability

- Consistency
- Transparency
- Targeting

(Better Regulation Task Force 1997)

2. *The Better Regulation Commission* was established in 2005 as a non-departmental body, sponsored by the Department for Business, Enterprise and Regulatory Reform (BERR) to provide independent advice to government, from business and other external stakeholders, about new regulatory proposals, and about the Government's overall regulatory performance. Its terms of reference were "To advise the Government on action to reduce unnecessary regulatory and administrative burdens, and ensure that regulation and its enforcement are proportionate, accountable, consistent, transparent and targeted" (Better Regulation Commission 2006). It was replaced by the Better Regulation Executive in 2008 (see below).
3. *The Local Better Regulation Office (LBRO)* was established in 2007. It was subsequently given a range of statutory duties and powers under the Regulatory Enforcement and Sanctions Act 2008 (Department for Business Innovation and Skills 2011). In 2009 it established and subsequently coordinated the Primary Authority scheme. Primary Authority enables businesses to form a statutory partnership with one local authority, or fire and rescue authority, which then provides robust and reliable advice for other local regulators to follow when carrying out inspections or addressing non-compliance. Agreements can cover broad or specific areas of environmental health, fire safety, licensing and trading standards legislation. The aim is to ensure that local regulation is consistent at a national level and sufficiently flexible to address local circumstances (Regulatory Delivery 2016a).The scheme was extended and simplified in 2016 and an updated handbook published (Regulatory Delivery 2016b). The LBRO was replaced by the BRDO in 2012 (see below).
4. *The Hampton Review.* In his final report (Hampton 2005) of his review for UK Government, Sir Phillip Hampton set out a vision for a regu-

latory system that targeted persistent offenders whilst supporting those businesses that want to comply. The report proposed:

- "reducing inspections where risks are low, but increasing them where necessary
- making much more use of advice, applying the principle of risk assessment
- substantially reducing the need for form-filling and other regulatory information requirements
- applying tougher and more consistent penalties where necessary
- reducing the number of regulators that businesses deal with from thirty-one to seven
- entrenching reform by requiring all new policies and regulations to consider enforcement, using existing structures wherever possible
- creating a business-led body at the centre of government to drive implementation of the recommendations and challenge departments on their regulatory performance".

Along with these specific recommendations, The Hampton Review set out some key principles that should be consistently applied throughout the regulatory system:

- "regulators, and the regulatory system as a whole, should use comprehensive risk assessment to concentrate resources on the areas that need them most
- regulators should be accountable for the efficiency and effectiveness of their activities, while remaining independent in the decisions they take
- no inspection should take place without a reason
- businesses should not have to give unnecessary information, nor give the same piece of information twice
- the few businesses that persistently break regulations should be identified quickly and face proportionate and meaningful sanctions
- regulators should provide authoritative, accessible advice easily and cheaply
- regulators should be of the right size and scope, and no new regulator should be created where an existing one can do the work

- regulators should recognize that a key element of their activity will be to allow, or even encourage, economic progress and only to intervene when there is a clear case for protection".

(Department for Business Innovation and Skills 2017)

5. *The Macrory Review of Sanctions.* The Hampton Review also highlighted the need to revisit the range of sanctions available to regulators. This was subsequently undertaken by Professor Richard Macrory in 2006 who recommended introducing a set of "civil sanctions" that would allow regulators to impose proportionate, flexible and meaningful sanctions in situations of regulatory non-compliance, as an alternative to criminal prosecution. The recommendations from these reviews were accepted by the UK government, and the conditions for granting and use of civil sanctioning powers were incorporated into the Regulatory and Enforcement Sanctions Act 2008 (RES Act). As this marked a significant change in enforcement practice, a proviso was imposed that, before a regulator can exercise these powers, ministers must be satisfied that the regulator will use the powers in accordance with the principles of good regulation, that is, that regulatory activities should be carried out in a way which is transparent, accountable, proportionate and consistent and should be targeted only at cases in which action is needed. These principles are set in statute by the Legislative and Regulatory Reform Act 2006 and exemplified by the Regulators' Code (Better Regulation Delivery Office 2015).
6. *The Better Regulation Executive.* As a result of the final recommendation of the Hampton Review above, the government created the Better Regulation Executive (BRE) to oversee the reduction of regulatory burdens on business, and hold government departments and regulators to account. This replaced the Better Regulation Commission as a joint unit of the former Department for Business, Innovation & Skills (BIS) and the Cabinet Office.
7. *The Hampton Implementation Reviews.* In November 2006, the government announced that "the National Audit Office would work with the Better Regulation Executive, regulators and business to develop a

process of external review of regulatory performance". The assessment process focussed on the extent to which regulators were performing in line with the Hampton principles and Macrory characteristics, and would encourage continuous improvement. Five major national regulators were to be assessed by the end of 2007 (Health and Safety Executive, Food Standards Agency, Financial Services Authority, Environment Agency (EA) and Office of Fair Trading) (National Audit Office 2007). The 2007 assessment of the Environment Agency (EA) is discussed below.

Better Regulation in the UK 2010–2016

The Coalition Government in 2010 put even more emphasis on better regulation. Better regulation was a central component of the Government's Economic Recovery Plan after 2010 in all departments. The prime minister's approach to the economy and to business included a desire to tackle regulation, as he expressed in his letter to cabinet ministers (see Box 3.4).

Specific bodies and initiatives in this period included:

8. *The Better Regulation Delivery Office* (BRDO). This was established in 2012 to replace the Local Better Regulation Office (LBRO). This was a directorate within BIS responsible for promoting the better delivery and enforcement of regulation. BRDO took over and further developed the Primary Authority scheme (Regulatory Delivery 2016b). "Good regulation", "regulatory reform", and "inspection reform" seemed to be preferred terms used by BRDO rather than

Box 3.4 Extract from the prime minister's letter to all cabinet ministers, 6 April 2011

"We need to tackle regulation with vigour to free businesses to compete and create jobs, and give people greater freedom and personal responsibility... I want us to be the first Government in modern history to leave office having reduced the overall burden of regulation, rather than increasing it" (Department for Business, Innovation and Skills 2015).

"better regulation". They described their purpose as "reforming regulatory policy, inspection reform and the reform of regulatory delivery", focussing on improving the way regulations are enforced, which includes inspection practice, culture and knowledge of regulators, risk analysis, business engagement and accountability (Better Regulation Delivery Office 2016). They set up an Inspection Reform Network and held international bi-annual conferences to promote regulatory reform. BRDO was merged with another body on 1 April 2016 to become "Regulatory Delivery".

9. *The Regulatory Policy Committee.* This is an advisory non-departmental public body, which scrutinises and validates the estimates that government departments make of the costs and benefits to business of regulations. It publishes its findings regularly (Regulatory Policy Committee 2016).
10. *The Government's approach to regulation, "Reducing Regulation made simple"*. This was published in December 2010. This was an important aspect of the Government's overarching objective of achieving sustainable economic growth. This set out the government's expectations of Parliament, Ministers, Civil Servants, Regulators and Business (Better Regulation Executive 2010) (See Box 3.5).

It also stressed the need to consider alternative approaches to regulation (see Box 3.6).

Box 3.5 The government's vision of a system of government for UK and European regulation

Parliament:	• thoroughly scrutinise both new regulatory proposals and old regulatory frameworks that are subject to review • challenge the impact of the regulatory framework and recommend actions to reduce the burden of ineffective regulation

Ministers:

- ensure that the government is only intervening when necessary and justified
- focus on identifying the most effective way to achieve desired policy outcomes
- resist the temptation for hasty regulation, even under intense media pressure
- exercise discipline in considering new regulatory measures, as any new costs must be offset by reductions elsewhere so the cumulative burden is reduced
- look at the cumulative impact of new EU measures
- encourage and challenge civil servants to actively explore creative, non-regulatory solutions to achieve desired policy outcomes, including in the EU
- encourage building alliances across Europe to increase the UK's effectiveness in achieving good regulatory results

Civil Servants policy-makers including lawyers and economists:

- develop appropriate incentives, skills and knowledge to actively explore and implement innovative, non-regulatory solutions to achieve desired policy outcomes
- are equipped with analytical tools to provide expert policy advice to ministers on non-regulatory options, including for implementing EU obligations
- develop effective policies that work constructively and imaginatively with businesses' and citizens' needs and circumstances
- enable businesses, individuals and other organisations to get involved in delivering the desired outcomes
- review more frequently whether policies are delivering the intended outcomes, and whether regulations can be simplified or removed to reduce regulatory burdens
- are equipped with the skills to work effectively across international boundaries—both at international and European level and with the devolved administrations

Enforcers of regulation:

- work with the grain of businesses' and other organisations' own incentives and processes, reducing oversight and inspection of organisations where effective self-regulatory systems and controls exist
- focus efforts on high-risk businesses, particularly those who deliberately seek to get an advantage over their competitors by breaking the law

Businesses, civil society groups and citizens:

- experience less intrusive, costly and prescriptive regulation
- can take greater personal responsibility for the way they deal with their obligations under the law in ways that suit their circumstances
- have more of a say in the way regulation is made and enforced

Box 3.6

Examples of alternative approaches to command and control regulation

Self-regulation: An approach initiated and undertaken by those whose behaviour is to be regulated. For example, an industry or profession might choose to develop and adopt its own code of practice promoting ethical conduct.

Examples:

- Unilateral codes of conduct
- Customer charter
- Unilateral sector codes
- Negotiated codes

Co-regulation: Similar to self-regulation but involves some degree of explicit government involvement. For example, an industry might work with government to develop a code of practice. The code would usually be enforced by the industry itself, or a professional organisation, rather than by the government.

Examples:

- Recognised codes
- Statutory codes
- Approved codes
- Voluntary agreements
- Trade Association codes approved by the Office of Fair Trading
- Accreditation and Standards

Information and education: Can be used to empower consumers to take their own informed decisions.

Examples:

- Inform, enhance consumer choice
- Independent recommendation schemes
- Ratings systems
- Labelling, disclosure.

Economic instruments: Can be used to modify behaviour by adjusting the economic incentives facing businesses and citizens. This approach allows individuals to make their own decisions, based on their estimates of whether the benefits of acting in a certain way justify the costs.

Examples:

- Taxes
- Subsidies
- Quotas and permits
- Vouchers
- Auctions
- Competition

No new intervention: In many instances, it might not be necessary for government to initiate new action at all. Regulation and its alternatives will almost always impose costs as well as generating benefits, so policy makers should think carefully about whether action by the government is required at all.

Examples:

- Use existing regulation
- Simplify or clarify existing regulation
- Improved enforcement of existing regulation
- Make legal remedies more accessible or cheaper
- Do nothing at all

(Better Regulation Executive. 2010)

11. *The Reducing Regulation Cabinet sub-committee (RRC)* has oversight of Government policy on regulation, including the Principles of Regulation (see Box 3.7). Scrutiny and clearance from the RRC is the central means by which better regulation concerns are reflected in the collective agreement process.
The RRC also clears the contents of the Statement of New Regulation (SNR): a six-monthly publication, setting out measures which will

Box 3.7 The UK Government's Principles of Regulation

The government will regulate to achieve its policy objectives only:

1. having demonstrated that satisfactory outcomes cannot be achieved by alternative, self-regulatory, or non-regulatory approaches
2. where analysis of the costs and benefits demonstrates that the regulatory approach is superior by a clear margin to alternative, self-regulatory or non-regulatory approaches
3. where the regulation and the enforcement framework can be implemented in a fashion which is demonstrably proportionate; accountable; consistent; transparent and targeted. There will be a general presumption that regulation should not impose costs and obligations on business, social enterprises, individuals and community groups unless a robust and compelling case has been made. The Government will adopt a One-in, One-out approach [now a One-in, Two-out approach].

(Department for Business, Innovation and Skills 2015)

come into force over the coming six months and reporting on progress under the One-in One-out and One-in Two-out rules.

12. *Business Impact Target.* The Government set a target, known as the Business Impact Target, to reduce the total cost of regulation for business by £10 billion between 2015 and 2020. The Better Regulation Executive is responsible for developing and implementing a framework for achieving these cost savings. Departments and regulators are responsible for delivering the cost savings to achieve the target through the regulatory decisions they make (House of Commons Committee of Public Accounts 2016).
13. *Red Tape Challenge (RTC).* This was a cross-government programme to review the stock of existing regulation. The default was that regulation should go unless it can be well defended. Following on from the RTC, each Government Department established "Better Regulation Units" and developed regulatory review measures. For example, the Better Regulation Unit in Defra was established and carried out the Smarter Environmental Regulation Review (SERR) which is described in detail in chapter 4. More details about the

impact of the RTC on *environmental* regulation are provided later in this chapter.

14. *The Regulators' Code*. The UK Government's Regulators' Code came into statutory force on 6 April 2014 and provides a clear, flexible and principles-based framework for how regulators should engage with those they regulate. The Code applies to most UK non-economic regulators, including local authorities, fire and rescue authorities and national regulators such as the EA who must have regarded to it when developing policies and procedures that guide their regulatory activities. There are six provisions of the Code:

    - Regulators should carry out their activities in a way that supports those they regulate to comply and grow;
    - Regulators should provide simple and straightforward ways to engage with those they regulate and hear their views;
    - Regulators should base their regulatory activities on risk;
    - Regulators should share information about compliance and risk;
    - Regulators should ensure clear information, guidance and advice are available to help those they regulate meet their responsibilities to comply; and
    - Regulators should ensure that their approach to their regulatory activities is transparent.

    (Better Regulation Delivery Office. 2014)

15. *Better Regulation Framework Manual*. In 2015, BIS published this practical guidance for UK Government Officials involved in policy development (Department for Business, Innovation and Skills 2015)
16. *Regulatory Delivery*. In April 2016, BRDO and the National Measurement and Regulation Office became a single directorate in BIS to focus on regulation and enforcement. The directorate is working to ensure that the way regulation is enforced is proportionate and risk based. It delivers functions such as Primary Authority, legal metrology and hallmarking policy, technical regulation and enforcement work (Regulatory Delivery 2016c).

Better Regulation in the European Union up to 2016

- In 2002, the EU Better Regulation Programme was established and was a first step in simplifying and improving EU legislation. It introduced obligatory impact assessments and stakeholder consultations for all new initiatives proposed by Commission.
- In 2005, the European Commission set out its position in a "Communication to the Council and the European Parliament" on "Better Regulation for Growth and Jobs in the European Union". This aimed to better design regulation so as to increase the benefits for citizens, and to reinforce the respect and the effectiveness of the rules, and to minimise economic costs—in line with the EU's proportionality and subsidiarity principles. The actions it proposed to achieve this were:
    - by further promoting the design and application of better regulation tools at the EU level, notably in so far as impact assessments and simplification are concerned.
    - by working more closely with Member States to ensure that better regulation principles are applied consistently throughout the EU by all regulators. As well as action at EU level, the communication emphasised the importance that transposition of EU legislation by the Member States and that national regulatory initiatives have a direct effect as well, not just on national administrations and on citizens but also on businesses, particularly SMEs, from across the Union.
    - by reinforcing the constructive dialogue between all regulators at the EU and national levels and with stakeholders (European Commission 2005).

The "Simplification Rolling Programme" was established and covered 164 simplification measures for 2005–2009 and became part of the annual Commission work programme.

- In 2007, the Commission proposed and the European Council agreed that burdens arising from EU legislation, including national rules implementing or transposing this legislation, should be reduced by 25% by 2012. It therefore established an "Action Programme for Reducing Administrative Burdens in the European Union". It also invited Member States to streamline their purely domestic legislation by setting "national targets of comparable ambition". It put the needs and concerns of small businesses at the very heart of the programme (European Commission 2010a). A "High Level Group of Independent Stakeholders on Administrative Burdens" was to advise the Commission on the implementation of the Action Programme. This "HLG" group was set up in 2007, chaired by Edward Stoiber, former prime minister of Bavaria consisting of 15 members selected on the basis of their expertise in better regulation and or policy areas covered by the action programme. The HLG adopted more than 30 opinions covering more than 300 suggestions on how to reduce burdens for businesses (High Level Group of Independent Stakeholders on Administrative Burdens 2011). By the end of the action programme, the Commission reached its target of cutting by 25% the administrative burden for businesses stemming from EU legislation (estimated annual savings €30.8 billion) (European Commission 2016f).

- In 2009, the Commission presented "Sectoral Reduction Plans" for each of the 13 areas covered by their Action Programme. These plans showed that the reduction measures already adopted, could lead to savings of €7.6 billion per year. That could become about €40 billion if the European Parliament and the Council backed the measures pending approval or under preparation. From seven Member States in 2006, all 27 had, by 2009, set ambitious targets for reducing burdens stemming from purely national rules (European Commission 2010b).
- In 2010, the Commission updated its 2005 Communication, with one entitled "Smart Regulation in The European Union" with the aim of ensuring that European laws benefit people and businesses. In this communication, the Commission set out plans to further improve the quality and relevance of EU legislation, in particular:
    - evaluating the impact of legislation during the whole policy cycle: when a policy is designed, when it is in place, and when it is revised
    - encourage Member States to apply "smart regulation" in their work
    - to increase the period of its public consultations from 2012 onwards so as to strengthen the voice of citizens and other stakeholders

(European Commission 2010b)

- In 2012, the Commission's Regulatory Fitness and Performance (REFIT) programme was established to make EU law simpler, and to reduce the costs of regulation while still achieving benefits. REFIT aims to keep EU law simple, remove unnecessary burdens and adapt existing legislation without compromising on policy objectives (European Commission 2016a). This is all part of the Commission's work on assessing the performance of the existing body of EU law and making changes where necessary to keep laws up to date.
- Since the appointment of Jean Claude Juncker as Commission president in 2014, the Commission made clear that it would break with the past and change the way the Commission works and sets its policies, by putting better regulation principles at the heart of its policy-making processes, to make sure its policies deliver better results for citizens, businesses and public authorities. It committed to be "big on the big things and small on the small things", by focussing action on those issues that really matter to the citizens and where European action is most necessary, and making sure that Member States take responsibility where national action is more appropriate (European Commission 2016b). Specifically, it strengthened REFIT, by creating more possibilities for stakeholders and EU countries to contribute. The REFIT platform (European Commission 2016g) chaired by the new First Vice-President Timmermans, collects suggestions (European Commission 2016f) and makes recommendations on how to simplify laws. David Cameron himself provided a list of suggestions of what the UK suggested removing. The state of play of REFIT initiatives for simplification and reducing regulatory burden is published in the REFIT scoreboard (European Commission 2016f). Evaluations and fitness checks are used to assess if EU laws, policies and funding programmes are delivering the expected results at minimum cost (European Commission 2016h).
- In its September 2016 Communication entitled "Better Regulation: Delivering better results for a stronger Union" (European Commission 2016b), the Commission produced a very useful summary of its various better regulation initiatives and progress made (see Box 3.8). For

Box 3.8 Better regulation initiatives of the European Commission in numbers

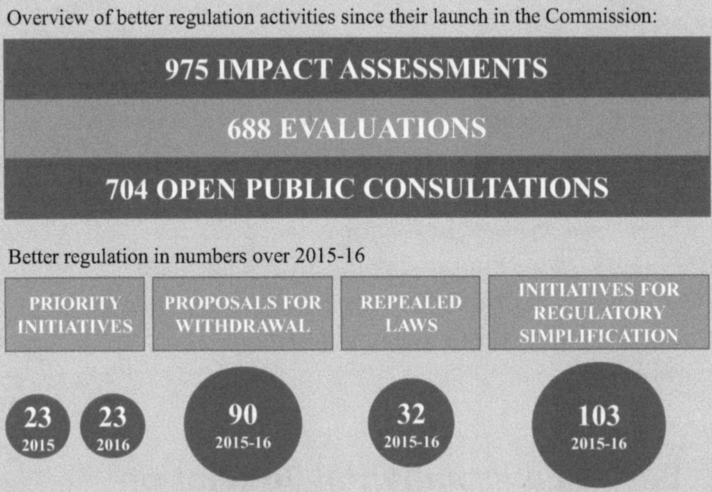

European Commission (2016b)

example, since the launch of the REFIT, almost 200 initiatives for burden reduction and simplification have been launched. This includes a reduction of up to 95% in the registration fees for SMEs required by chemicals legislation (REACH), and a reduction in administrative costs associated with compliance monitoring and reporting by approximately €345,000–€460,000 per year, for binding annual greenhouse gas emission reductions by Member States from 2021 to 2030. This was complemented by further simplifications in monitoring and reporting across energy and climate-related legislation.

- The "Inter-institutional Agreement on Better Law-making" reached by the European Parliament, the Council, and the European Commission, entered into force in 2016 (European Commission 2016d). This clarified the way in which these bodies would work together on various matters, mostly concerned with better regulation, and according to the Commission, marked a significant step forward in the culture of better regulation.

As a result of the above better regulation initiatives, the work of the Commission is now very much more about evaluating, reviewing, simplifying and improving its current stock of legislation and the way it is implemented rather than drafting of new legislation. Indeed, in the Environmental arena, the 7th European Environmental Action Programme (European Commission 2014) had very little in the way of new legislation but emphasised the following four so-called "enablers" which will help Europe deliver on its environmental goals:

- better *implementation* of legislation
- better *information* by improving the knowledge base
- more and wiser *investment* for environment and climate policy
- full *integration* of environmental requirements and considerations into other policies

3.4 Better *Environmental* Regulation

The better *environmental* regulation initiatives over the last 20 years or so have generally been undertaken within the *wide-ranging* better framework covered above. Having said that, the environmental better regulation initiatives in the UK have often led the way for other regulatory regimes. One example is the risk-based inspection system OPRA[3] which was developed in the first half of the 1990s by Her Majesty's Inspectorate of Pollution (HMIP) and later implemented by the EA. This was perhaps the first published risk-based inspection methodology in the UK, since when other regulatory regimes have developed similar systems and BRDO has developed a common methodology for use by national regulators and local authority regulatory services (Better Regulation Delivery Office 2012).

Better Environmental Regulation Initiatives in the UK

As might be expected, these were led by the EA in the 1990s. At that time, other environmental regulators with smaller resources, such as the

Scottish Environment Protection Agency (SEPA) and the similar Northern Ireland Environment and Heritage Service often followed the EA in England and Wales and modified EA systems for their own use. However, over more recent years, the environmental regulators in Scotland, Northern Ireland and Wales have developed better regulation initiatives at a similar pace to the EA in England.

The Environment Agency's Better Environmental Regulation

The EA has developed several better regulation approaches since its formation in 1996. In 2005 it published its regulatory approach (Environment Agency 2005).

These were summarised in 2006 as shown in Box 3.9.

Box 3.9 Four Environment Agency tools which fit together to make up its compliance assessment process

Operator and pollution risk appraisal (OPRA) scheme

The OPRA scheme assesses the environmental risk of an activity. It assesses the hazards associated with an activity and how well they are being managed. The assessment provides a risk-rating or profile which EA uses as part of its compliance assessment process. The OPRA score for an activity determines how much EA charges businesses for regulating an activity.

Compliance assessment plans (CAP)

Compliance Assessment Plans are used to match EA's regulatory effort and available resource to the risk profile that OPRA has given the activity.

Methodology for assessing compliance (MAC)

The Methodology for Assessing Compliance is a guide for EA staff undertaking all types of compliance assessment activities including audits, inspections and sampling. The guidance enables a consistent and transparent approach to be taken to compliance assessment across regulatory regimes. It describes what to do to assess compliance and how to report the findings.

Compliance classification scheme (CCS)

The Compliance Classification Scheme provides consistency across different regulatory regimes in the reporting of non-compliance with permit conditions and the action the EA takes. The information from the CCS contributes to an activity's OPRA risk rating or profile.

Diagram showing how the above four Environment Agency tools all fit together:

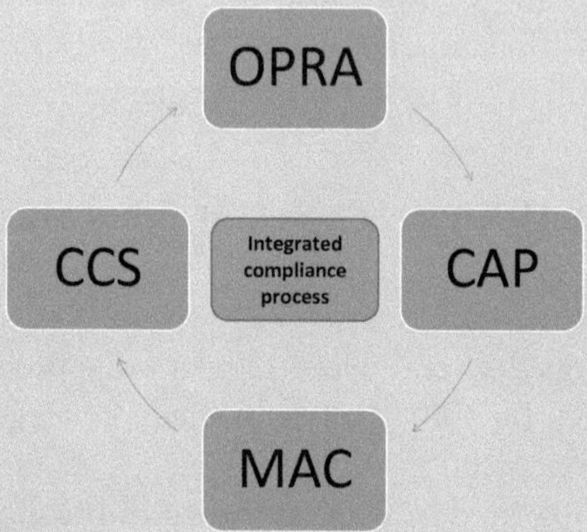

Box 3.9.1 Four agency tools (Environment Agency 2006)

The Hampton Implementation Review team examined the EA and concluded, in its report in March 2008, that the EA had taken significant steps to implement the Hampton agenda, although there was further progress to be made in translating the strong lead shown at senior levels to the day-to-day experience of individual businesses. It found that better regulation was certainly in the language of the organisation, but was not yet embedded throughout its culture (Better Regulation Executive and National Audit Office 2008). The specific findings are shown in Box 3.10.

Since the Hampton Implementation Review and the change of government, the EA has developed its regulatory approach further from the model described in Box 3.9. They are now placing more emphasis on achieving outcomes and on gathering intelligence, and helping businesses to manage compliance themselves through internal environmental management systems. This is allowing the EA to focus its inspection work on those sites and activities where it has particular concerns.

The EA is also developing new ways of working that are improving alignment with business sectors. For example, as part of their Future

Box 3.10 Better regulation in the Environment Agency in 2008 (as described in the findings of the Hampton Implementation Review)

The EA has had a strong commitment to better regulation over a number of years

The EA Board and Chief Executive are clearly committed to modern regulation, and are taking the lead in embedding this within the organisation. This is reflected in the EA's work to re-examine various aspects of its operations, its move towards a more risk-based approach to regulating, and its efforts to influence debate in Europe.

The EA has made encouraging progress in implementing the Hampton principles

The EA has taken forward a number of initiatives to support its modern regulation programme. These include, for example: the instigation of a Regulatory Scrutiny Panel to provide internal challenge to legislative and policy development; the Operator and Pollution Risk Appraisal (OPRA) as a tool for assessing environmental risk; and the recent implementation of the agricultural waste regulations with a high degree of consultation with the farming sector. There is scope for OPRA to be used more effectively to incentivise compliance and to inform risk-based interventions.

The EA is making progress in reducing the burden of regulations on business

The EA has made considerable efforts to rationalise and simplify the complex EU Directive-driven regulatory environment, from the viewpoint of regulated industries. Notable initiatives so far include the removal of some activities from regulation, such as some water abstraction activities and some low-risk hazardous waste activities. Other examples include:

- NetRegs, a web-based source of environmental guidance for businesses—although relatively few businesses actually use this resource;
- the National Customer Contact Centre (NCCC), which provides a single telephone point of contact and advice for business and public inquiries;
- the Environmental Permitting Programme (due to be implemented shortly), which will further build on the Pollution Prevention and Control (PPC) covering the discharge of damaging substances into the environment, and also incorporate waste management licensing into one simplified regime; and
- the Integrated Regulation programme (due to be implemented shortly), software which will bring together all the EA's online regulatory activities, including permit applications and payments. Further work is required regarding the rationale for undertaking inspections and requiring businesses to collect monitoring data, as these continue to place burdens on businesses whilst often having no clear link to environmental outcomes.

The EA is taking a lead in Europe in driving the better regulation agenda. The EA is using its influence to work with the European Commission and agencies in other Member States to help ensure that new regulations are in line with the principles of better regulation.

There is a general agreement amongst business stakeholders that the EA uses prosecutions in a fair and proportionate way. There is also recognition that, in some areas, the EA lacks the range of sanctions that would enable it to maximise its effectiveness in implementing regulations.

The EA has made notable efforts to improve the quality of its written advice, forms and publications

The EA has invested time and resources into producing new guidance and forms that are accessible and easy to read or complete. Recent external documents are written in plain English and the number of documents is being reduced. There are some good examples of guidance being produced in consultation with businesses, although there remains a significant legacy of written publications which are somewhat older and do not meet the EA's new standards. This and other initiatives would have greater impact if there was a more strategic approach to coordinate the different strands of advice and guidance provision: the EA reported that it has started to develop such an approach.(Better Regulation Executive and National Audit Office 2008)

Approach to Regulation (FAR) programme, it has reorganised regulation of sites according to key business sectors, allowing staff with expertise in specific sectors to provide a more tailored engagement. Working in this way allows the EA to gather and use sector-specific intelligence more effectively at all stages of the regulatory cycle, helping to make the right intervention at the right place and time (Environment Agency 2014). The sector-based approach is discussed in more detail in chapter 5.

The current EA model for regulation as of 2014 updates the previous one in the diagram in Box 3.9 and makes a few small changes to reflect the changes in emphasis. It is summarised in Box 3.11. No doubt this will be amended and updated as the EA continues to develop its approach, while at the same time facing cuts to its budget. For example, in October 2016, it suspended its "definition of waste panel" following budget cuts. This service advised businesses on creating new products out of their waste, providing advice on the EU End of Waste Regulations. The panel consisted of agency experts who helped businesses with this process (Chartered Institution of Wastes Management 2016).

Box 3.11 The Environment Agency's model for environmental regulation

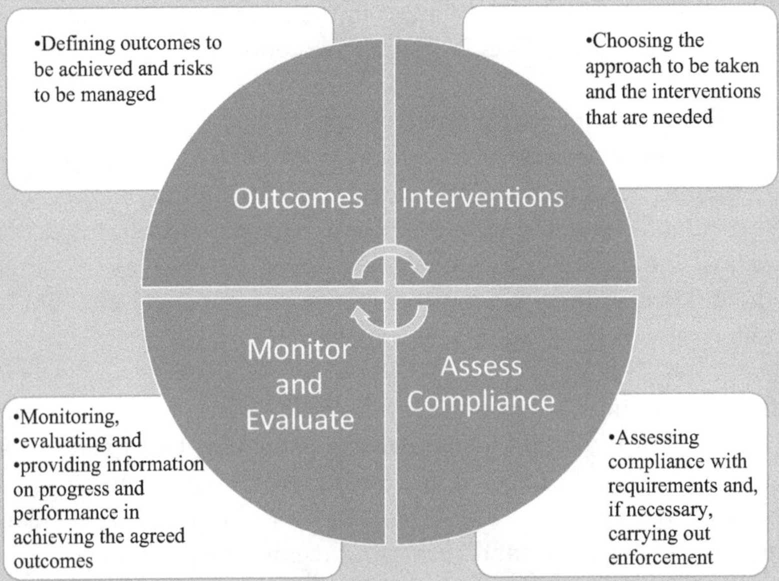

Simplifying and Streamlining Environmental Legislation

A few reviews have taken place of the overall suite of environmental legislation applying to businesses to establish how legislation as a whole is operating in any given area. Reviews have covered, for example: to what extent the legislation appears to be a coherent integrated whole; whether it is consistent in its terminology, philosophies, descriptions of offences and penalties, reporting requirements, appeals; and whether it can be easily interpreted by users (Department for Environment, Food and Rural Affairs 2013a). The purpose of such reviews has generally been to simplify, streamline, integrate and make the whole body of legislation easier to understand and to apply by businesses and environmental regulators.

One such review which lasted from 1998 to 2002 was undertaken by the Department of Environment Transport and the Regions and the EA (Department for Environment, Food and Rural Affairs 2003).

The review addressed several areas in the form of a critical analysis of existing regimes, procedures and legislative controls. It focussed on:

- the procedures or mechanisms currently used by the agencies, for example an opportunity to consider a consolidated consent, authorisation or licence, in place of individual consents, and so on.
- the simplification and rationalisation of control regimes, for example radioactive and heavy industrial processes.
- the differences in legislation (between water, air, waste etc.) and the different ways in which these are operated.

The review did not recommend wholesale revision of environmental legislation, but it did lead to another successful joint Defra and EA programme, the "environmental permitting regime programme" which replaced several pieces of legislation and procedures and terminologies with a single environmental permitting regime (Department for Food, Environment and Rural Affairs 2013). The Environmental Permitting Regulations 2010 that implemented the new regime, were generally applauded for addressing integration problems overall (UK Environmental Law Association 2012).

Notwithstanding the success of the environmental permitting regimes, it was recognised that there was still a number of areas that could be improved. For example:

- An examination of the provision for environmental appeals uncovered that they appear in over 60 pieces of legislation relating to the environment. Appeals go to a wide range of different bodies including the High Court, Magistrates Court, the Planning Inspectorate, and different government departments (Macrory 2011). The report of this examination concluded that "The system lacks common procedure and intelligibility … There is little in the way of underlying principle in choice of the appeal body … Over the years we have developed a system of environmental appeals which is complex and confusing".

- An examination of the effectiveness of UK environmental legislation in terms of whether there are problems of lack of coherence, integration and transparency concluded that the environmental permitting regime was "a significant success in this respect, by integrating the administration of a number of regulatory regimes in England and Wales". However, it recognised that "problems of legislative coherence, integration and transparency not only impede the effectiveness of environmental legislation; they may also undermine the rule of law" (UK Environmental Law Association 2012). The report made 18 recommendations including the following three which called for improvements in coherence and integration:
  - governments should consolidate legislation more routinely
  - governments should provide an on-line single port of call for information about all relevant environmental legislation and associated guidance on a particular topic
  - further work should be done to consider the potential role that environmental principles might play in UK environmental law to address problems of legislative coherence, integration and transparency.

Following on from the above, Defra included a review of environmental legislation as one element of its Smarter Environmental Regulation Review in Defra (Department for the Environment Food and Rural Affairs (Defra) 2012). The terms of reference said England's environmental law had evolved in a "piecemeal way", and it was tasked with proposing a new long-term direction and framework. It examined models in other countries, such as Wales' two new sustainability laws, and work undertaken in Germany to reformulate its environmental legislation. It asked whether you could come up with a legislative outcome that was better for the environment and better from a business and regulatory efficiency point of view, and the answer was pretty strongly "yes" from all the sectors consulted. A report was produced within Defra but at the time of writing has not been published (Kaminski 2016).

In its report on manufacturers' views of progress on Defra's regulatory reform agenda, the Engineering Employers' Federation said: "there is a strong appetite amongst manufacturers for deeper, legislative reform of environmental policy to bring together the layers of legislation that have developed over the years into a more coherent and understandable structure" and "Government should commit to wide-sweeping reform of environmental legislation to rationalise and consolidate the existing stock, but without compromising the levels of environmental protection afforded by it" (Engineering Employers' Federation 2015).

Northern Ireland has taken several measures in its Better Environmental Regulation Act (Northern Ireland Assembly 2016) to streamline and integrate its environmental legislation. For example:

- Creation of an integrated environmental permitting regime, which will allow the Department to issue an integrated permit that, where possible, will cover all environmental controls that relate to activities being undertaken at a site
- Rationalising powers of entry and associated powers such as inspection and investigation

In Scotland, a number of better environmental regulation initiatives have introduced more streamlined, integrated and consistent approaches, for example:

- The single authorisation framework which will bring together all the permissioning arrangements for SEPA's four main regulatory (water, waste, radioactive substances and Pollution Prevention and Control) into a single permissioning structure and under a single standardised procedure (Scottish Government 2017).
- SEPA's regulatory charging scheme replaces five existing schemes covering waste, waste exemptions, radioactive substances, Pollution Prevention and Control and water discharges and abstractions with a single, risk-based scheme. This makes SEPA's charging simpler, more consistent and transparent (Scottish Government 2016b).
- New Environmental Enforcement Framework. Historically, the options for dealing with environmental offences had been limited and inconsistent and focussed more on the offence rather than changing the behaviour. SEPA now has a suite of new enforcement measures and Scotland's courts have new powers (Scottish Government 2016a).

In Wales, there have been several initiatives to streamline and integrate environmental legislation and its application, since the formation of Natural Resources Wales. The Wellbeing of Future Generations Act (Welsh Government 2015) and the Environment (Wales) Act (Welsh Government 2016) both include numerous provisions in this regard.

Better Regulation Delivery Office Reviews of Regulators Against the Principles of Good Regulation

BRDO carry out regular reviews against the Principles of Good Regulation. In addition to providing useful feedback to regulators on how to improve, assurance that a regulator is following the principles of good regulation is a prerequisite for it to be granted civil sanctioning powers.

In 2015, following one of its regular reviews of regulators, BRDO published its report on its review of Natural Resources Wales against the Principles of Good Regulation.

The review team noted the following key areas where NRW is performing particularly well:

- NRW has made efforts to embed all six provisions of the Regulators' Code in its policy, culture and practice.
- NRW shows ambition to develop a new and different regulatory approach which recognise its diverse role, business expectations and focus on outcomes.
- NRW recognises the value of engagement and has put in place a variety of mechanisms to ensure those they regulate, or who are impacted by regulation have the opportunity to share their views.
- The NRW approach to working with the hydropower sector is noteworthy and an excellent example of regulatory co-production.
- Transparency is a key strength of NRW.

(Better Regulation Delivery Office 2015)

DEFRA's Better Environmental Regulation Initiatives

The most significant better regulation work of Defra recently has been the Smarter Environmental Regulation Review. This is covered in chapter 4.

A central feature of Defra's approach to better environmental regulation is to drive higher compliance by focussing on the users and simplifying and improving regulation and regulatory requirements so that more businesses know what they have to do. It established a Better Regulation Unit to produce evidence and reform Defra's regulations in accordance with Government Better Regulation Policy. Its key work on Better Regulation is published on the Gov.UK website (Department for Environment, Food and Rural Affairs 2013d).

One key aspect of that work was dealing with the Red Tape Challenge (RTC) so far as it affected environmental regulations. "The responses received as part of the RTC made clear that neither the public nor the industry wants to see environmental or other protections watered down. Nevertheless, it was equally clear that the accumulation of rules, guidance, reporting, and inspections makes it sometimes impossible to put the rules into practice. The biggest short term gains could be made from making what to do in practice much simpler and clearer" (Lockhart-Mummery 2015). As a result, Defra established the Smarter Environmental Regulation Review, which is described in detail in chapter 4.

"The Red Tape Challenge included a consideration of 278 pieces of legislation relating to the environment. From one side (the business camp) came forth gasps of exasperation at the myriad of overlapping or apparently unnecessary requirements that were hard to comprehend and costly and time-consuming to comply with; from the other (the green camp), cries of terror that this might be a cunning wheeze to do away with important environmental protections". (Oliver, Smarter and wiser? An update on Defra's Smarter Environmental Regulation Review 2014).

However, "in a statement about the Red Tape Challenge dated 25 March 2015, Defra Minister George Eustice announced that by May the Department would have made '650 legislative reforms' which would reduce the number of regulations in force by over 20% compared to May

2010 … This covers all of Defra's legislation, not just its environment laws but includes many measures that relate to the environment … They mainly consist of reforms to tidy up what had become a rather messy legislative sprawl … This has been done partly through revoking or repealing moribund legislation such as the water classification schemes that have been superseded by the Water Framework Directive. There has also been some impressive consolidation and simplification of requirements that were spread across several statutory instruments, or that had been subject to numerous amendments. For example, the Environmental Liability Directive—formerly transposed by a total of twelve different statutory instruments—has been re-implemented by the Environmental Damage Regulations 2015 (which also introduce new measures concerning offshore oil and gas). Similarly, the Nitrates Directives—formerly transposed through six separate sets of regulations—has been re-implemented by the Nitrate Pollution Prevention Regulations 2015 … The Red Tape Challenge then so far as the environment was concerned did not do away with environmentally protective measures or downgrade standards by characterising the requirements as costly red tape" (Oliver, Environmental regulation—stripping back bureaucracy or protection? 2015).

Defra committed to developing an Instrument Selection Guide for Policy makers (Department for Environment, Food and Rural Affairs 2013b) and published a very useful "draft" Instrument Selection Guidance at the end of that year (Department for Environment, Food and Rural Affairs 2013c).

Better Environmental Regulation Initiatives in the EU

The European Network for the Implementation of Environmental Law (IMPEL) have undertaken several projects on better regulation. It established a "Better Regulation Cluster" to commission projects and publish reports to help European environmental regulators ensure a more effective application of environmental law. For example:

- It produced a "Better Regulation Checklist" to assess practicability and enforceability of legislation (IMPEL, European Union Network for the Implementation and Enforcement of Environmental Law, 2010).
- It examined and reported on "Complementary Approaches to Environmental Inspections for Ensuring Compliance" (IMPEL, European Union Network for the Implementation and Enforcement of Environmental Law, 2012).
- It developed a toolkit for choosing appropriate interventions alongside inspections to ensure environmental compliance and achieve environmental outcomes (IMPEL, European Union Network for the Implementation and Enforcement of Environmental Law, 2014).

The Make it Work Project (MiW) is an initiative by The Netherlands (Ministry of Infrastructure and the Environment), the UK (Defra), Germany (Federal Ministry of Environment, Nature Conservation, Building and Nuclear Safety), Sweden (Ministry of Environment and Energy) and Czech Republic (Ministry of the Environment). The aim of the project is to identify concrete opportunities to improve the quality of EU environmental law and thus help to achieve the benefits associated with the law while delivering a more level playing field across the EU. In particular, it aims at establishing a more coherent and consistent framework for the EU environmental acquis through developing drafting principles on the use of cross-cutting instruments and procedures in EU environmental directives and regulations. MiW aims at delivering environmental outcomes more efficiently and effectively, without lowering existing protection standards. At the time of writing it has produced "drafting principles" for compliance assurance, and for environmental reporting and the Commission Fitness Check on Environmental Monitoring and Reporting (Institute for European Environmental Policy 2016).

The 7th EU Environmental Action programme has a big focus on implementation rather than (as in previous action programmes) developing new legislation. Examples include:

- integrating environmental and climate-related conditionalities and incentives in policy initiatives, including reviews and reforms of existing policy, as well as new initiatives, at Union and Member State level;
- carrying out ex-ante assessments of the environmental, social, and economic impacts of policy initiatives at appropriate Union and Member State level to ensure their coherence and effectiveness;
- using ex-post evaluation information relating to experience with implementation of the environment acquis in order to improve its consistency and coherence.

- (European Commission 2014)

3.5 Critics of Better Regulation

There is undoubtedly a certain amount of opposition to better regulation on political grounds, that it is seen as just about removing regulation, and is looking at regulation only from the aspect of "burdens" on business rather than benefits to the environment, people and indeed to business themselves. This opposition is certainly not comforted by rhetoric such as "*one in two out* or *the chance to rip up some of the 21,000 rules and regulations that are getting in your way*" (Department for Business Innovation and Skills 2011b). One example of such opposition entitled "better regulation for whom" challenges the labour, coalition and conservative government's better regulation programmes (Tombs 2016) (see Box 3.12).

Tombs also makes the point that regulation is widely derided, that the politics of anti-regulation have been overlain by the economics of austerity, and austerity has particularly impacted upon regulation and enforcement at the level of local authorities (through reductions in numbers of enforcement officers).

While it is undeniable that budgets to regulators (national as well as local) have been reduced, many examples of better regulation initiatives described in this chapter are not just about politically driven deregula-

Box 3.12 A criticism of government's better environmental regulation agenda

"The rate of inspection and enforcement actions for environmental health, food safety and hygiene, and health and safety have all been falling. In the case of health and safety inspections by local authorities, for instance, the average business can now expect to be visited only once in every 20 years. This is not just a problem of infrequent inspections and lax enforcement. In the name of cutting red tape, governments of all political persuasions have, for over a decade, undermined independent and effective business regulation. Budget cuts under the austerity programme have compounded the problem. So too have moves to outsource and privatise regulatory and enforcement activity. Private companies are increasingly involved in 'regulating' themselves. Taken together these changes may ... mark the beginning of the end of the state's commitment to, and ability to deliver, social protection". Source: Tombs (2016)

tion, but have been designed to make things easier for regulated and regulators, while at the same time safeguarding environmental protections. Indeed, if regulators had seen budget cuts without any better regulation initiatives, it would perhaps be even harder to maintain compliance and environmental standards. Exactly how much any initiative has saved business (or not) and protected the environment (or not) is hard to say and ex-post evaluations are extremely thin on the ground.

Another criticism of the government's better regulation agenda, regarding environmental regulation in particular has come from the Aldersgate Group (Aldersgate Group 2011). Their view is summarised in Box 3.13:

Meanwhile environmental lawyers, through the UK Environmental Law Association have had concerns about better regulation (see Box 3.14).

From businesses' point of view, many better regulation initiatives have been welcomed, but there are calls for further work on coherence and integration. See Box 3.15.

Box 3.13 Assessment of Government's regulatory reform

While reducing outdated, excessive and burdensome measures are welcome, this must not be at the expense of the vital role that regulation plays in correcting market failures, promoting fairness and protecting the environment. Regulatory reform should be primarily concerned with the effective achievement of outcomes and maximising innovation potential. Costs must be minimised but this should not be the only guiding principle.

Given that regulation is one of the few means of stimulating the economic recovery to take a more sustainable path, an overly rigid regulatory reform framework risks damaging competitiveness. A mind-set of "best-in, bad-out", rather than "one-in, one-out" that takes a whole system approach, would, for example, tailor needs more effectively to specific challenges.

Through streamlining legislation and adopting a smarter approach to implementation, it is possible to achieve greener outcomes, reduce regulatory burdens and make business in the UK more competitive and attractive. But this goal means focussing on the desired outcomes and holistic analysis of the benefits of the regulatory and non-regulatory interventions required. Choices based solely on merit and value must not be constrained by arbitrary targets on cutting red tape or achieving short-term cost benefits. Source: Aldersgate Group (2011)

Box 3.14 Environmental lawyers' views on Defra's regulatory reforms

The UK Environmental Law Association (UKELA) recognised that Defra's better regulation reforms (Department for the Environment, Food and Rural Affairs 2014) were timely and had the potential to address many of the issues that UKELA raised in its own report (UK Environmental Law Association 2012). However, they recognised the "danger that the ambition behind the 'once-in-a-generation opportunity to set a clearer direction for environmental legislation' is thwarted by forces outside Defra's control" (Oliver, Smarter and wiser? An update on Defra's Smarter Environmental Regulation Review 2014).

Regarding the Red Tape Challenge, UKELA commented that "Defra's initial follow-up was somewhat underwhelming. Mainly this involved identifying moribund legislation that could be repealed, and some candidates for consolidation: essentially raking up fallen leaves and trimming the hedges, rather than re-landscaping the regulatory garden" (Oliver, Smarter and wiser? An update on Defra's Smarter Environmental Regulation Review 2014).

Regarding simplification of guidance UKELA commented "Rationalisation as a means of avoiding inconsistency and confusion is certainly a welcome step. However, there is a real risk that the drive to reduce massively the overall volume could mean that useful and important detailed or technical guidance is axed, causing uncertainty about compliance requirements. Further, the drive for simplicity and clarity seems to be resulting in guidance documents that merely describe what legislation requires, without guiding on matters of policy and interpretation. Greater uncertainty about such matters could result in more cases being litigated: not a particularly 'smart' result" (Oliver, Smarter and wiser? An update on Defra's Smarter Environmental Regulation Review 2014).

"UKELA is ...concerned that the scale of Defra's 'cull' may be leading to the loss of useful guidance. In the course of the review, some guidance documents in areas such as waste and WEEE have been disappearing completely from Gov.UK with no explanation, and no apparent replacement. This has been causing problems both for practitioners advising on past issues who need to access the guidance that was in force at the time, as well as those advising on current issues who are faced with uncertainty as to how to interpret and apply regulatory requirements....". "Useful guidance may also be lost in the process of developing new documents that are 'simpler, quicker and clearer to use and understand'. There are fears that the new documents will include little by way of actual guidance on matters such as how to interpret key terms and regulators' policy approaches to implementing legislation" (Oliver, Environmental regulation—stripping back bureaucracy or protection? 2015).

Box 3.15 Manufacturers' views of progress on Defra's regulatory reform agenda

- Environmental regulatory reform has the potential to cut inefficiency in both government and industry and can help to enhance the perception of UK competitiveness without compromising environmental protection.
- Manufacturers support recent efforts by Defra to simplify data reporting and guidance. Improvements here are considered to be important to business performance.
- However, manufacturers are not yet feeling the benefit of the efforts made thus far. The ultimate goal should be a single point of access for guidance for each core area and one single data reporting system for all environmental data.
- Furthermore, there is a strong appetite amongst manufacturers for deeper, legislative reform of environmental policy to bring together the layers of legislation that have developed over the years into a more coherent and understandable structure.
- To be effective, the UK's push for deregulation must take place both domestically and within Europe.

Source (Engineering Employers' Federation 2015)

3.6 Latest Developments in Regulatory Strategies

Many environmental regulators have published how they regulate and how they have incorporated better regulation principles; for example, the EA (Environment Agency 2014) set out its regulatory principles, ways of working, its model for environmental regulation (see Box 3.11), and its regulatory approaches and instruments.

Perhaps the most recent and forward-looking regulatory strategy though was published by SEPA in 2016 entitled "One Planet Prosperity" (Scottish Environment Protection Agency 2016a). It includes several elements of "better regulation" as described above (although it does not use the term itself once). It rightly recognises how the Regulatory Reform Act (Scottish Government 2014) gave SEPA the opportunity to create what it claims will be "one of the first environmental regulatory systems in the world that is suited to the challenges of the twenty-first century". This includes new enforcement powers and integrated authorisation and charging frameworks mentioned above as well as SEPA's "Waste to resources framework" (Scottish Environment Protection Agency 2016b). It describes very neatly how and why the role of environmental regulators has changed since the end of the twentieth century, recognising the changes in business and the environment and the other influences on businesses environmental behaviour which it describes as follows:

- consumer demand for environmental credentials
- investor requirements for environmental performance
- supply-chain requirements for environmental performance
- assessment by external ratings bodies
- trade association membership standards
- expectations of potential employees about environmental performance
- social scrutiny (e.g. residents, NGOs) and via social media (e.g. Twitter)

The key challenge for SEPA is described as combining the things they can do to influence the behaviour of a business, with all the other influences on the behaviour of that business (see Box 3.16). This is seen by

Box 3.16 Regulators Influence Map

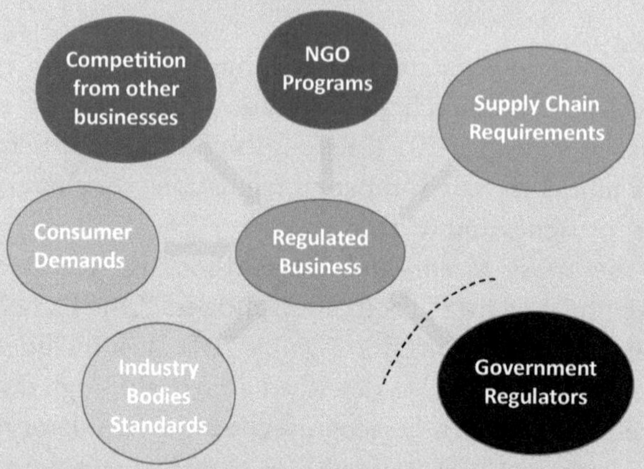

Source: Scottish Environment Protection Agency (2016a)

SEPA as the most effective way to deliver full compliance, and help as many businesses as possible move beyond compliance.

3.7 What Might the Future Hold for "Better" Environmental Regulation

Much has been achieved in reforming regulation over the past 20 years or so in Europe and the UK and so-called better regulation initiatives have played a large part in that. Whether this was for good or ill depends on your point of view, but it obviously limits further scope for regulatory reform.

Better regulation has gone a long way to reduce burdens and simplify things *for business.* This is as true for other aspects of regulation as for environmental regulation. At the same time though, so far as environmental regulation is concerned at least, many of the better regulation initiatives also aimed to make regulation better *for achieving environmental outcomes.* That is not to say that there is not more to do on improving the way that regulation works to make it simpler for business and regulators and the public to understand and implement regulation. Efforts to streamline, harmonise and integrate environmental legislation fall into that category. The removal of European legislative barriers to this after Brexit may help in that regard.

However, the author's view (maybe as much as in hope as in expectation) is that the future is likely to see environmental regulators put even more focus on making regulation better for the environment while at the same time working within reducing budgets.

The key to this, I think, is improving our understanding of the influences on business to behave in an environmentally sound way (or the reverse) and only making regulatory interventions when the influence of regulation is actually going to make a difference. If not, then regulators and governments will be looking to apply other interventions (directly or through other "actors") to influence business behaviour. SEPA's "Regulators Influence Map" in Box 3.16 helps to explain, and chapter. 8 covers recent work on designing and choosing interventions and selecting actors to deliver them so I will not discuss that in any more detail here. However, put simply, the way that this can be done is by recognising all the motivations on business, and then working out the regulator's role and its interventions within the context of that bigger picture. This means focussing on how can regulators and government bolster the positive incentives and counter the negative incentives delivered by those other influences on business environmental behaviour. Naturally this will require a shift of skills and activities of regulators and governments towards increased analysis and intelligence of each particular set of circumstances in which businesses operate and in which they create actual or potential environmental harm.

Of course, such developments are not free of risk, in particular, the risk that any choice of interventions (and actors to deliver them) will not actually improve environmental performance of business. So, alongside any such initiatives there needs to be a much better understanding of "what works in what circumstances and why".

So, to summarise, the areas that I think are likely to occupy those concerned with reforming environmental regulation in the future and how it is applied are as follows:

- Wholesale rationalisation of the body of environmental law. Progress in this regard has been good in some countries such as Scotland, Wales and the Netherlands. So far as England is concerned, much of the

preparatory work has been done (Kaminski 2016). However, EU legislation restricted Defra's opportunities for deeper, legislative reforms of environmental policy to bring together the layers of legislation that have developed over the years into a more coherent and understandable structure. Perhaps after the UK leaves the EU such barriers to rationalisation may disappear.
- A better understanding of the many influences on businesses' environmental performance alongside that of the environmental regulators role (see Box 3.16).
- Development and application of a range of interventions by regulators and other "delivery agents" in the optimum mix for each particular set of circumstances to improve business environmental behaviours so as to meet legal requirements and environmental objectives.
- More evaluation of *what works and why,* so that policy makers and regulators have more evidence to help them choose appropriate instruments and delivery agents, according to circumstances, to achieve their environmental objectives.

3.8 Summary and Conclusions

Much has been achieved in reforming regulation over the past 20 years or so in Europe and the UK and the so-called better regulation initiatives have played a large part in that. Much of the emphasis of such initiatives were about removing burdens and improving regulation from the perspectives of businesses that were subject to regulation.

Reform of *environmental* regulation has arisen partly under the umbrella of the broader better regulation agenda but has also been instigated by initiatives of environmental regulators such as the EA and SEPA in order to improve the effectiveness of regulation in improving the environment.

In the future, the areas that the author thinks are likely to occupy those concerned with reforming environmental regulation and how it is applied are as follows:

- Wholesale rationalisation of the body of environmental law to bring together the layers of legislation that have developed over the years into a more coherent and understandable structure

- A better understanding of the many influences on businesses' environmental performance alongside that of the environmental regulators role
- Development and application of a range of interventions by regulators and other delivery agents according to circumstances to improve business environmental behaviours
- More evaluation of *what works and why*, so that policy makers and regulators have more evidence to help them choose appropriate instruments and delivery agents, according to circumstances, to achieve their environmental objectives.

Notes

1. Business in this context would include individuals only so much as they are self-employed or one-person businesses. Farmers might come into that category as might self-employed haulage contractors transporting waste for example.
2. This begs the questions "did any of these initiatives actually make things better?" and indeed "better for who?" Unfortunately, there is not a great deal of evidence on the effectiveness of regulation (better or otherwise) but the author has tried to include any where known.
3. OPRA is covered in more detail in chapter 1.

References

Aldersgate Group. 2011. *Dealing with Deficits. Best Value Regulation to Reduce Our Environmental and Financial Debts*. London: Aldersgate Group.

Better Regulation Commission. 2006. *About Us*. May 4. Accessed from www.brc.gov.uk; https://web.archive.org/web/20061013175118/http:/www.brc.gov.uk/about_us/

Better Regulation Delivery Office. 2012. *Proposals for Developing a Common Approach to Risk Assessment*. London: Department for Business Innovation and Skills.

Better Regulation Delivery Office. 2014. *Regulators' Code*. London: Department for Business Innovation and Skills.

Better Regulation Delivery Office. 2015. *A Review of Natural Resources Wales Against the Principles of Good Regulation*. London: Department for Business Innovation and Skills.

Better Regulation Delivery Office. 2016. *BRDO LinkedIn Pages.* October 14. Accessed from LinkedIn: https://www.linkedin.com/company/better-regulation-delivery-office

Better Regulation Executive. 2010. *Reducing Regulation Made Simple.* London: HM Government.

Better Regulation Executive and National Audit Office. 2008. *Effective Inspection and Enforcement: Implementing the Hampton Vision in the Environment Agency.* London: Better Regulation Executive.

Better Regulation Task Force. 1997. *Principles of Good Regulation.* http://webarchive.nationalarchives.gov.uk/20100407162704/http:/archive.cabinetoffice.gov.uk/brc/upload/assets/www.brc.gov.uk/principlesleaflet.pdf.

Chartered Institution of Wastes Management. 2016. Environment Agency Suspends Waste Reuse Advisory Service. *CIWM Journal* Online October 2016, 1.

Department for Business Innovation and Skills. 2011a. *The Future of the Local Better Regulation Office and extending the benefits of the Primary Authority Scheme.* London: HM Government.

Department for Business, Innovation and Skills. 2015. *Better Regulation Framework Manual. Practical Guidance for UK Government Officials.*

Department for Business Innovation and Skills. 2017. *Assessing Our Regulatory System—The Hampton Review.* January 11. http://webarchive.nationalarchives.gov.uk/:http://webarchive.nationalarchives.gov.uk/20100407162704/http:/archive.cabinetoffice.gov.uk/brc/upload/assets/www.brc.gov.uk/principlesleaflet.pdf

Department for Business Innovation and Skills. 2011b. *Red Tape Challenge Kicks Off.* April 7. Accessed from gov.uk: https://www.gov.uk/government/news/red-tape-challenge-kicks-off

Department for Environment, Food and Rural Affairs. 2003. *Review of Legislation relating to Integration Within the Environment Agency and Scottish Environment Protection Agency.* London: Department for Environment, Food and Rural Affairs.

Department for Environment, Food and Rural Affairs. 2013a. *20 Ways to Influence Business Behaviour. A Short Guide to Instrument Selection for Policy Makers and Regulators.* London: Department for Environment, Food and Rural Affairs.

Department for Environment, Food and Rural Affairs. 2013b. *Better Evidence for Regulatory Reform: Rapid Evidence Appraisal Number 5: Environmental Policy Integration.* London: Defra.

Department for Environment, Food and Rural Affairs. 2013c. *Better Regulation Strategy Unit Evidence Plan 2013/14–2107/18.* London: Department for Environment, Food and Rural Affairs.

Department for Environment, Food and Rural Affairs. 2013d. *Better Regulation in Defra*. August 7. Accessed from Gov.Uk: https://www.gov.uk/government/collections/better-regulation-in-defra

Department for Food, Environment and Rural Affairs. 2013. *Environmental Permitting Guidance Core Guidance For the Environmental Permitting (England and Wales) Regulations 2010*. London: Department for Food, Environment and Rural Affairs.

Department for the Environment, Food and Rural Affairs. 2014. *Terms of Reference for Smarter Environmental Legislation: Phase 2 of the Smarter Environmental Regulation Review*. London: Defra.

Department for the Environment Food and Rural Affairs (Defra). (2012). *Terms of Reference for the Smarter Environmental Regulation Review.* http://webarchive.nationalarchives.gov.uk/

Engineering Employers' Federation. 2015. *Green Tape: Manufacturers' Views of Progress on Defra's Regulatory Reform Agenda*. London: Engineering Employers' Federation.

Environment Agency. 2005. *Delivering for the Environment: A 21st Century Approach to Regulation*. London: Environment Agency.

Environment Agency. 2006. *Guide to Modern Regulation*. London: Environment Agency.

Environment Agency. 2014. *Regulating for People, the Environment and Growth*. London: Environment Agency.

European Commission. (2005). *Better Regulation for Growth and Jobs in the European Union*. Brussels: European Commission.

European Commission. (2010a). *Action Programme for Reducing Administrative Burdens in the European Union*. Brussels: European Commission.

European Commission. 2010b. *Smart Regulation in the European Union*. Brussels: European Commission.

European Commission. 2014. *7th European Environmental Action Programme*. Brussels: European Commission.

European Commission. 2016a. *Better Regulation Why and How.* Accessed from European Commission: http://ec.europa.eu/info/law/law-making-process/better-regulation-why-and-how_en

European Commission. 2016b. *Better Regulation: Delivering Better Results for a Stronger Union*. Brussels: European Commission.

European Commission. 2016c. *European Commission.* Accessed from REFIT—Making EU Law Simpler and Less Costly: http://ec.europa.eu/info/law/law-making-process/overview-law-making-process/evaluating-and-improving-existing-laws/reducing-burdens-and-simplifying-law/refit-making-eu-law-simpler-and-less-costly_en#documents

European Commission. 2016d. *Interinstitutional Agreement Between the European Parliament, the Council of the European Union and the European Commission on Better Law-Making*. Brussels: European Commission.

European Commission. 2016f. *REFIT—Making EU Law Simpler and Less Costly.* Accessed from European Commission: http://ec.europa.eu/info/law/law-making-process/overview-law-making-process/evaluating-and-improving-existing-laws/reducing-burdens-and-simplifying-law/refit-making-eu-law-simpler-and-less-costly_en#documents

European Commission. 2016g. *REFIT Platform.* Accessed from European Commission: http://ec.europa.eu/info/law/law-making-process/overview-law-making-process/evaluating-and-improving-existing-laws/reducing-burdens-and-simplifying-law/refit-making-eu-law-simpler-and-less-costly_en#documents

European Commission. 2016h. *Evaluating Laws, Policies and Funding Programmes.* Accessed from European Commission: http://ec.europa.eu/info/law/law-making-process/overview-law-making-process/evaluating-and-improving-existing-laws/evaluating_en

European Union. 2015. *Guidelines on Evaluation and Fitness Checks.* May 19. Accessed from European Union: http://ec.europa.eu/smart-regulation/guidelines/ug_chap6_en.htm

Hampton, P. 2005. *Reducing Administrative Burdens. Effective Inspection and Enforcement.* London: HM Treasury.

High Level Group of Independent Stakeholders on Administrative Burdens. 2011. *Europe Can do Better. Report on Best Practice in European States to Implement EU Legislation in the Least Burdensome Way.* Brussels: European Commission.

HM Government. 2008. *Site Waste Management Plans Regulations 2008.* Accessed from legislation.gov.uk: http://www.legislation.gov.uk/uksi/2008/314/contents/made

HM Government. 2013. *The Environmental Noise, Site Waste Management Plans and Spreadable Fats etc. (Revocations and Amendments) Regulations 2013.* Accessed from legislation.gov.uk: http://www.legislation.gov.uk/uksi/2013/2854/regulation/2/made

House of Commons Committee of Public Accounts. 2016. *Better Regulation Eighteenth Report of Session 2016–17.*

IMPEL (European Union Network for the Implementation and Enforcement of Environmental Law). 2010. *Better Regulation Checklist.* Brussels: IMPEL.

IMPEL. 2012. *Exploring the Use and Effectiveness of Complementary Approaches to Environmental Inspections for Ensuring Compliance.* Brussels: IMPEL.

IMPEL. 2014. *Choosing Appropriate Interventions Alongside Inspections to Ensure Environmental Compliance and Achieve Environmental Outcomes. Final Report.* Brussels: IMPEL.

Institute for European Environmental Policy. 2016. *What is Make it Work.* Accessed from Institute for European Environmental Policy: http://www.ieep.eu/work-areas/environmental-governance/better-regulation/make-it-work/home/

Kaminski, I. 2016. *Buried DEFRA Policy Work 'Should Be Unearthed' to Develop New Green Law.* Ends, July 27.

Lockhart-Mummery, E. 2015. *Clearing the Thicket Blog.* September 10. Accessed from Civil Service Quarterly Blog: https://quarterly.blog.gov.uk/2015/09/10/clearing-the-thicket/#comments

Macrory, R. 2011. *Consistency and Effectiveness. Strengthening the new Environmental Tribunal.* London: Centre for Law and the Environment, University College London.

National Audit Office. 2007. *Hampton Implementation Reviews. Guidance for Review Teams.* London: NAO.

Northern Ireland Assembly. 2016. *Environmental Better Regulation Act (Northern Ireland) 2016.* Belfast: Northern Ireland: The Stationary Office.

Oliver, R. 2014. Smarter and Wiser? An Update on Defra's Smarter Environmental Regulation Review. *UKELA "e-law" Newsletter. Issue 84*, 31–33.

Oliver, R. 2015. *Environmental Regulation—Stripping Back Bureaucracy or Protection?* July 20. Accessed from LexisNexis: https://www.lexisnexis.com/uk/lexispsl/commercial/document/412012/5GGJ-3K81-DYW7-W0BH-00000-00/Environmental_regulation_stripping_back_bureaucracy_or_protection_

Regulatory Delivery. 2016a. *Local Regulation: Primary Authority.* September 19. Accessed from Gov.UK: https://www.gov.uk/guidance/local-regulation-primary-authority

Regulatory Delivery. 2016b. *Primary Authority Handbook.* London: Department for Business, Energy and Industrial Strategy.

Regulatory Delivery. 2016c. *Regulatory Delivery News Issues 1.* London: Regulatory Delivery; A Directorate of the Department of Business Innovation and Skills.

Regulatory Policy Committee. 2016. *Costs and Benefits of New Regulation.* March 10. Accessed from Gov.uk: https://www.gov.uk/government/publications/costs-and-benefits-of-new-regulation-regulatory-policy-committee-opinions-since-may-2015/costs-and-benefits-of-new-regulation

Scottish Environment Protection Agency. 2016a. *One Planet Prosperity—Our Regulatory Strategy.* Scottish Environment Protection Agency.

Scottish Environment Protection Agency. (2016b). *One Planet Prosperity—Waste to Resources Framework.* Scottish Environment Protection Agency.

Scottish Government. 2014. *Regulatory Reform (Scotland) Act 2014.* The Stationary Office.

Scottish Government. 2016a. *New Environmental Enforcement Measures.* April 29. Accessed from gov.scot: http://www.gov.scot/Topics/Environment/waste-and-pollution/BER/EnforcementMeasures

Scottish Government. 2016b. *SEPA's New Charging Scheme.* April 29. Accessed from gov.scot: http://www.gov.scot/Topics/Environment/waste-and-pollution/BER/ChargingScheme

Scottish Government. 2017. *Consultation on Proposals for an Integrated Authorisation Framework.* Edinburgh: Scottish Government.

Tombs, S. 2016. *Better Regulation for Whom. Briefing 14 April 2016.* London: Centre for Crime and Justice Studies.

UK Environmental Law Association. 2012. *The State of UK Environmental Legislation in 2011–12: Is There a Case for Legislative Reform?* London: UK Environmental Law Association, King's College London and Cardiff University.

Welsh Government. 2015. *Well-being of Future Generations (Wales) Act 2015.* Wales: The Stationary Office.

Welsh Government. 2016. *Environment (Wales) Act.* Wales: The Stationary Office.

Chris Booth worked for almost 40 years as an environmental professional in the field of environmental regulation in the UK as a regulator, as a policy practitioner, as a research manager and latterly as an independent consultant. He also worked in a water company, acting as an interface with the environmental regulators. Chris developed specific expertise in complementary and alternative approaches to regulation that policy makers and regulators can use to improve the environmental performance of businesses. This includes understanding the circumstances and drivers of businesses, which determine which interventions might work best to improve their environmental performance. Chris retired as an environmental professional in 2016 to pursue other interests.

4

Steps Towards Radically Smarter Regulation in the UK (2012–2017)

Edward Lockhart-Mummery

4.1 A New Period of Reform Begins

This chapter builds on Chap. 3 and plots the emergence of a new wave of reforms to environmental regulation from the early 2010s. It includes personal perspectives from the author's experience in leading and advising on reforms across a number of government bodies in England and Scotland. Each of these reforms tends to focus on one or more of the following:

- *Integrating environmental objectives across government policy and the economy*. Smarter environmental legislation advocates using the best-place levers to achieve outcomes and not narrowly relying on regulation within environment ministries. The Welsh Environment Act requires a wide range of government bodies to take account of environmental objectives (covered in Chap. 3). Scotland's One Planet

E. Lockhart-Mummery (✉)
Director Resource Decisions Ltd, London, UK

J. Foreman (ed.), *Developments in Environmental Regulation*, Palgrave Studies in Environmental Policy and Regulation, DOI 10.1007/978-3-319-61937-8_4

Prosperity reforms illustrate harnessing all factors that influence businesses to achieve results, such as consumer and supply chain demands, instead of relying entirely on 'command and control' methods.

- *Prescribing outcomes not method or process*. This normally involves replacing detailed and often quickly changing rules with requirements to meet stable long-term outcomes giving business both the room to innovate and confidence to invest. Smarter Environmental Legislation advocates rewiring the policy framework to focus on outcomes, Scotland's One Planet Prosperity sets about moving to outcome focussed permits and Smarter Guidance, in simply setting out the legal requirements and removing prescription, helps to provide as much flexibility as possible within the law in meeting outcomes.
- *Consolidating multiple environmental objectives*. Environmental regulations have typically evolved to tackle single issues resulting in multiple delivery chains which are administratively intensive. It can also be complex and confusing for users. The Welsh Environment Act, Northern Ireland Better Regulation Act (covered in Chap. 3), One Planet Prosperity, smarter data and smarter environmental legislation all involve consolidating multiple objectives into single more coherent regulatory systems.
- *Designing regulatory arrangements around their users*. This mainly includes simplifying regulations by using the latest digital technologies and methods that makes it radically simpler, cheaper and quicker to work out what to do and follow processes, designing out duplication at the same time.
- *Redefining government's role as regulator*. While the concept of market failure provides a clear rationale for government intervention in the environment, there are no clear or consistent principles about the scope and role for regulators in the market. This has created wide variance across policies and regions and has crowded out market solutions. The Smarter Guidance project articulates a clearer role in core command and control regulation, while One Planet Prosperity illustrates how partnership working can help harness wider influences to achieve more ambitious outcomes.

These early reforms have created a solid body of practical experience of what works and provides a platform for much deeper reform to integrate environmental policy objectives into the economy in a way that achieves the same or better environmental results, while making it easier to do business and invest in the future of our country.

4.2 The Development of Environmental Regulation

The development of environmental policy follows broadly the same pattern anywhere in the world. Citizens and public interest groups become concerned about specific issues such as polluted air, water or soil, waste dumps, chemical hazards or decline in species and habitats. Rules are introduced to regulate how activities are managed. Rules and the administrative systems that support them develop and grow incrementally.

European Union (EU) environmental regulation and legislation, which accounted for roughly 80% (by cost) of England's environmental regulations (Department for the Environment, Food and Rural Affairs 2011a), had been remarkably effective at solving twentieth-century environmental problems reducing the worst forms of damage (e.g. water quality, air quality, recycling rates).

4.3 Early 2010s: A Changing Context for Environmental Regulation

However, by the early 2010s, several pre-conditions for reforming UK environmental regulation were converging:

- *The EU environmental framework was largely complete.* Since 1973, UK environmental policy had been dominated by EU environmental action programmes and lawmaking. In 2013, the 7th Environmental Action Programme (European Commission 2013) recognised that the environmental acquis now covered the major environmental issues.

There would be less focus on new laws and more on how best to achieve the objectives.

- *Environmental challenges are becoming more pressing.* Recent analysis (European Environment Agency 2015; Global Footrpint Network 2017) shows that over-use of planetary resources not only in the UK and EU and increasingly worldwide is undermining our ability to sustain social progress and economic growth. Specific problems in the UK include 40,000 early deaths each year from poor air quality and over 30% of natural assets in decline. The nature of the challenges has also changed from largely point source and acute problems to chronic challenges of diffuse sources of pollution, over-use of natural resources and major environmental challenges, such as climate change.
- *Social trends are likely to add further pressure.* Global population is expected to increase from 7 billion to 10 billion by 2020. The share of populations living in urban areas is expected to continue to rise. Middle-class consumption is going up (European Environment Agency 2015).
- *The environmental management context has changed.* The culture and capability within businesses for managing environmental issues have changed beyond recognition from previous decades. Environmental performance is now driven by many other influences such as customers, supply chains, investors, local communities and the public and internal cost reduction and risk management pressures. Many companies have their own environmental management and reporting systems. This all changes the role for government regulation.
- *Policy can be made and implemented in new ways.* Environmental regulation and regulators largely deal in the exchange of information—about what you need to do, how to do it, whether you've done it and what impact that's had. The rise of digital technologies and methods has huge potential to unrecognisably disrupt and transform regulatory design and implementation. Maturing understanding and attitudes to the environment also create room for using open and co-designed policy-making methods that are gaining popularity in the UK and internationally. At the same time, government is moving away from managing outcomes through delivery chains to creating the conditions in which performance will improve.

- *Business increasingly needs to compete in global markets.* This affects what business needs from environmental regulation in various ways. It increases pressure to reduce production costs including regulatory costs. Policy stability is important to attract long-term investment. The appetite to innovate domestically and then sell solutions into international markets, for example, in infrastructure solutions, often depends on technology-neutral policies.

4.4 Growing Recognition of the Need for Change

Against this changing context, there is growing consensus that regulation needs to be reinvented if it is to succeed in the twenty-first century. Recent research and assessments (Department for the Environment, Food and Rural Affairs 2011b, 2013, 2014a; HM Government 2011, 2017; Lawton 2010; EU Member States 2017; UK Environmental Law Association 2012) shed light on a number of structural problems which will increasingly constrain our ability to achieve a healthy environment, dynamic economy and resilient society. These, largely, result from the way environmental policies have evolved over many decades and the fact that various pressures, at both EU and national level, have prevented serious reform until now. Some of the structural flaws with the today's policy environment include:

- *Disproportionate focus on regulating certain industries rather than integrating in all relevant policy.* Environment ministries and the EU's DG environment have low leverage over relevant policy domains such as infrastructure, industrial policy, transport, housing, planning, technology, funding and fiscal policy. To achieve their objectives, they therefore over-rely on making detailed regulations in the limited areas where they have levers. Take the short-term hikes on manufacturing air emissions limits rather than addressing transport systems and infrastructure.
- *Imposing constraints rather than giving responsibility.* Environmental policy does not generally give people positive responsibility for the environment.

Instead, it imposes constraints on actions. This encourages both a business mindset where 'I can do anything unless told not to' and policymakers to make rules covering anything that might cause harm. Inevitably, this leads to inconsistency in how we treat different problems.

- *Focus on detailed rules rather than outcomes*. Specifying technical requirements rather than outcomes also constrains the scope for using innovative solutions that regulators haven't thought of. It reduces the scope for trade-offs between competing environmental objectives to make sure maximum benefit is achieved overall. It reduces credibility of environmental policy where rules have lost sight of the purpose, for example, where disproportionate action is taken to protect small numbers of species individuals rather than improve total population health. It creates duplication between regulators' requirements to assure compliance and companies' own internal systems for managing risk.
- *Rules that are fragmented, duplicating and complex, instead of designed around users*. Incremental, issue-based evolution of policy and latterly transposition deadlines for EU directives have left us with a landscape of 1233 laws (The National Archives 2017) and even more complex arrangements to implement them.[1] This makes managing almost any regulated activity enormously more expensive and complex than necessary. For example, one large waste company has over 500 pieces of environmental legislation on its legislative inventory (Department for Environment, Food and Rural Affairs, Expected in 2017). Despite this, that same company already has corporate goals which take them far beyond the aims of the regulation.
- *Policy instability and unpredictability*. Almost any operational change with positive effects on the environment requires time to implement, some level of investment and often partnership or dependencies between different organisations. This holds whether it is investment in modern industrial plant, implementation of new business models or land management changes, and relies on a certain level of stability in the regulatory environment. However, both because of the number scale of rules and because rules are subject to unpredictable and short-term political decision-making, environmental regulation is often considered volatile undermining investment confidence.

- *Weak enforcement.* Simple, proportionate and robust enforcement is essential as a backstop to any system of rules. Yet environmental regulators are constrained by a complex and inconsistent landscape of powers and sanctions, rules about how it should enforce, what funds it can dedicate to enforcement and the disproportionately low level of financial sanctions compared to other regulatory areas such as competition law. Weak systems for ensuring that government and other public bodies comply with their legal duties and responsibilities. No national institutional arrangements—other than Ombudsmen dealing with maladministration or reliance on third parties bringing time-consuming and expensive judicial review actions.

4.5 England: The Economic Trigger

In England, the global financial crisis was then the catalyst for recent reforms. The Conservative-Liberal Democrat coalition government that came to power in the UK in 2010 in the wake of the 2007–2008 crisis brought with it a renewed focus on reforming regulation (HM Government 2010). Reducing unnecessary regulatory burdens was seen as one of the most effective ways to stimulate and help rebalance the UK economy away from an overdependence on financial services (HM Treasury 2011).

Debate quickly sparked around environmental regulation against the concern that reform would mean removing substantive protections rather than moving to genuinely better ways to achieve the intended outcomes. A progressive consensus emerged amongst many stakeholder groups about the nature of progressive reform to improve the regulatory landscape from both environmental and business perspectives (Aldersgate Group 2011). However, pressure groups continued to be concerned about the government's intentions, often provoked by comments, for example, about the Climate Change Act and the Habitats Directive. This arguably reduced public confidence and the space for sustained cross-sectoral collaboration that is essential to achieve lasting reform.

4.6 England: The Red Tape Challenge

The coalition government's main process to generate reform ideas from 2010 to 2015 was the Red Tape Challenge (HM Government 2017). This Cabinet Office-led "crowdsourcing" process involved scrutinising regulations theme by theme for reforms. Environmental regulations were included across a number of themes including environment, maritime, agricultural, housing and construction, hospitality, food and drink and manufacturing. The process involved:

1. publishing a list of regulations online, theme by theme
2. asking the public for views on which should be scrapped or improved
3. asking civil servants to put forward proposals including any further ideas
4. "star chambers" where cabinet office ministers challenged civil servants' proposals

Having reviewed each regulation one by one, the star chamber came up with a list of largely minor changes. It concluded that the core goals of regulations were sensible and that levels of protection should be retained, but that there was a big opportunity to improve environmental regulation as a whole. The Department for Environment Food and Rural Affairs (Defra), therefore, submitted radical proposals for cross-cutting reform to improve the regulatory framework including to overhaul guidance, reporting, inspections and legislation.

4.7 England: The Smarter Environmental Regulation Review

This review, launched in July 2012, took a user-centred approach to reform. The Executive Summary highlights the problem (Department for the Environment, Food and Rural Affairs 2013):

> Over recent decades, successive governments have built a framework of environmental regulation that has transformed the way we treat and value

> our environment in England. However, this framework has evolved in a piecemeal way and it now consists of hundreds of laws, guidance documents and procedures. Evidence from the Red Tape Challenge (RTC) exercise showed that the framework can appear fragmented, overlapping, inconsistent and complex. For some businesses this may act as a barrier both to effective compliance with their environmental obligations and to growth.

The review's core aim was to find ways to radically reduce burdens, while increasing the effectiveness of regulations. The review started by understanding the current situation through:

- comprehensive mapping of the existing stock of environmental legislation together with the associated guidance, information obligations and inspections
- eleven detailed sector case studies to understand how businesses comply with regulations on the ground and the issues faced

A first phase of recommendations was then made to reform guidance and information obligations. This was on the basis that quick progress could be made without major changes in legislation. A 'backcasting' approach was used, working with businesses in the 11 sectors and other leading environmental experts, to establish how regulatory arrangements should ideally work from a user perspective to achieve the desired outcomes. This led to a destination statement (Department for the Environment Food and Rural Affairs 2014c) for environmental regulation to help shape all future reforms (Fig. 4.1).

Proposals were then made for the wholesale reform of the system guidance and information reporting (see the smarter guidance and smarter data sections below). This was a departure from previous improvement exercises which tended to make more minor changes to individual areas, documents or measures in isolation. A second phase to develop proposals for reform of the underlying legislation was then taken up as the Smarter Environmental Legislation project (see the smarter environmental legislation section). Defra also initiated a farm visits project to reform all inspections to farm businesses.

Fig. 4.1 Destination statement for environmental regulation

4.8 England: The Smarter Guidance Reforms

These reforms rewrote all environmental guidance around what users need with a reduction in the volume (of pages and words) by over 80%.

Why Reform Guidance?

Evidence shows that people primarily look to guidance to work out what they need to do and rarely refer to the law itself. Therefore, ministers concluded that in terms of simplifying existing requirements, reforming guidance would have more impact and could be done quicker than reforming the law.

Environmental guidance had evolved in a piecemeal way as both legislation and the administrative structures to implement it had evolved. By 2012, Defra and other departments and organisations dealing with environmental rules had over 8000 documents in force totalling over 120,000 pages. Five new documents were published every working day. Documents were spread over multiple websites, there was no common format or style for guidance, no means to avoid overlap and no system for keeping it up to date. On average, the required reading age for environmental guidance was 21, whereas the national average reading age is nine (Department for the Environment, Food and Rural Affairs 2013).

The estimated costs to business of using the guidance was over £200 m pa and the government costs of producing and maintaining this guidance was over £20 m pa (Department for the Environment, Food and Rural Affairs 2014b). The evidence showed that awareness and understanding of what to do was low, and that the guidance often led to high uncertainty. There was also anecdotal evidence that this caused a barrier to business growth.

Objectives and Principles for Reform

The task was to make guidance: quick to find and read, simple to understand and clear so people know what they must do. Ministers also set a target to reduce volume by over 80%. This was to help communicate the scale of the change to be undertaken, both within organisations and to businesses and the public.

The reforms were guided by some core principles:

- *Design guidance for the user*. This means producing guidance around what people are doing, giving them exactly the information they need to carry out their task. The starting point is to work out who the users are, what they are trying to do and why: see Fig. 4.2. The method was then to develop a content plan to give precisely the information people need.

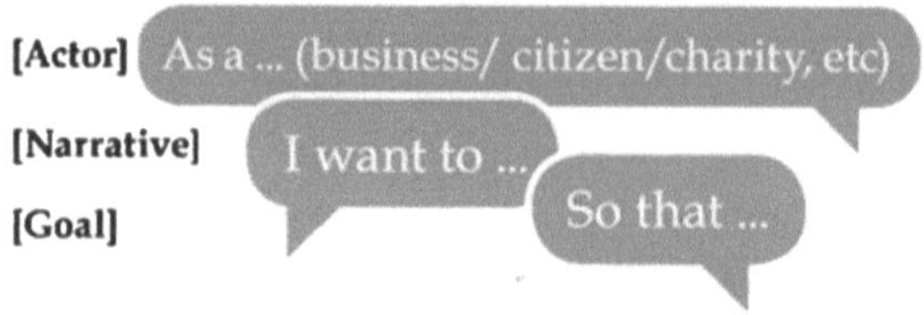

Fig. 4.2 The anatomy of a 'user need' (Lockhart-Mummery, Clearing the Thicket 2015)

- This is in stark contrast to producing guidance that explains what legislative instruments mean, or what government would like people to do or how to interact specifically with one organisation. Implicitly it involves creating a single version of what people need to do. Behind the scenes, this means many organisations working together on shared content. Designing for the user also aligned with another objective of Ministers which was to give businesses a 'safe harbour', that is, that if they followed they should be confident in doing what the law required. Previously, even comprehensive guidance based around the legislation were normally accompanied by waivers to the effect that even if you followed all the guidance, regulators might still take enforcement action against you.

- Designing for the user was the most fundamental change from the status quo. It was greatly assisted by the creation of GOV.UK—the single government website in England. Many of the methods for designing user-centred government guidance were, therefore, developed jointly between Defra and the Government Digital Service.
- *Use data and insights on what users are looking for.* Web analytics, including, for example, web searches, page views and bounce rates, reveal what people are looking for when they go online. It is striking how these information sources alone often quickly show how guidance content should be designed. For example, a review of searches for pet passports shows that the vast majority of people needing to get a pet passport want simply to know when they need a passport, how to apply, how much it costs, what to do if they lose their passport, and how to change or renew the passport. Yet much of this information was missing or obscured in previous versions of guidance leading to confusion and high volumes of customer calls.

- *Do what only government can do*. There were previously no clear principles for when government produced guidance and when other organisations such as professional bodies, trade associations, lawyers, academics or NGOs should produce guidance. Organisations complained that government's approach was random and unpredictable, preventing those otherwise well placed to do so from producing guidance for their members or clients. Defra concluded that government was uniquely placed and should consistently produce guidance to cover (Department for Environment, Food and Rural Affairs 2013):
  - *the law*—to explain what people must do
  - *definitions or basic advice*—where the law is not clear
  - *processes*—for example, for grants, permits and emergency situations
  - *how we take decisions*—for example, on permits, grants or enforcement

This excluded 'good practice' guidance explaining different operational methods for managing the environment. This type of guidance had previously been responsible for around 36% of the volume of Defra's agencies' guidance. Government good practice guidance had, however, caused a number of problems. First, businesses were unclear what was required by law and what was simply that body's view of good practice. Second, government rarely updated its good practice documents, thereby risking prescribing practices that were no longer good practice. Third, it reduced the motivation for private and third sector bodies to develop innovative new best practices and publish them.

Nevertheless, this was a radical and, for some people, a controversial change. There are two main types of good practice guidance:

1. *Guidance that sets out the methods that can be used to comply with legal requirements*. The debate was as follows. People felt that government should set out the methods because otherwise people will not know what to do. Government argued that it should clarify anything that they unambiguously require or expect to achieve compliance, but that otherwise they should leave it for others to develop and publish the methods. Defra also assessed where there were transitional gaps in

information and capability and worked with third-party organisations to ensure information remained available to those who needed it.

2. *Guidance that explains how environmental risks not covered by legislation should be managed.* The debate was as follows: people felt that, despite not having any legal status, government good practice guidance would encourage people and businesses to adopt management measures. Government argued that, in the absence of legislation, encouraging industry to produce its own guidance with the support of its membership would be more effective than government-published good practice guidance (Department for Environment, Food and Rural Affairs 2017b).

- *Write plain language, consistently.* Businesses had often found previous guidance difficult to read and understand, for example, because unambiguous or unfamiliar terms were used or because there was little consistency of writing style or terminology even across related documents. Defra, therefore, used the Government Digital Service's 'style guide' which is designed to make content easy to understand and consistent.
- *Write for the web.* Most people read guidance online. Searching for and reading content online is very different from using printed text. Reasons include the following: people can locate the text they need based on search terms, content can be linked and bundled up on the basis of what individual users are looking for and the way people read content has evolved. See, for example, one eye-tracking study recorded how over 200 users looked at thousands of websites and found that people read web content in an F-shape: one horizontal movement across the upper part of the content, then a second horizontal movement further down the page and then a vertical movement down the content's left side (Nielsen Norman Group 2006). This shows how content needs to be structured to make sure readers properly absorb your content. See the F-shaped image in https://www.nngroup.com/articles/f-shaped-pattern-reading-web-content/.

Thus, the content needs to be designed in different ways. Both 'content architecture' and 'search engine optimisation' are essential. Content architecture means, for example, breaking content down into the right chunks so that people only read what is relevant to them and organising content to put information in the order people need to know it. Search engine optimisation means positioning the terms that people search for correctly so that search engines find them.

What the Reforms Did?

The reforms were done using open policy-making principles to get useful public input as efficiently as possible. This involved seeking views and sharing detailed plans for new content and for archiving existing content via a dedicated reform website (Department for Environment, Food and Rural Affairs 2017a) and absorbing comments in real time. The main phase of guidance reforms took two years between 2013 and 2015. They covered all Defra and its agencies' guidance and environmental guidance issued by other departments including the department for Business Innovation and Skills (BIS), the department of Energy and Climate Change (DECC)[2] and the Department for Communities and their agencies.

To illustrate the results, let's take a closer look at how batteries waste was reformed (Fig. 4.3).

As with other subjects, guidance had evolved over a period of time as new statutory instruments were introduced, each organisation involved explained their arrangements and new documents were published to correct perceived failures of the existing body of guidance (Department for the Environment, Food and Rural Affairs 2013). This led, by 2012, to a total of 32 documents, 340 pages across six government websites. This meant that even the job of identifying which documents to read was challenging. In fact, there are four main activities users do involving batteries: importing or exporting, manufacturing or processing, selling or disposing of them. Having established what information users need for each, using data and user feedback, the team was able to write one document for each user need with on average five pages. This means that whatever you are doing, you can immediately Google and find the right document and quickly and confidently read what you need.

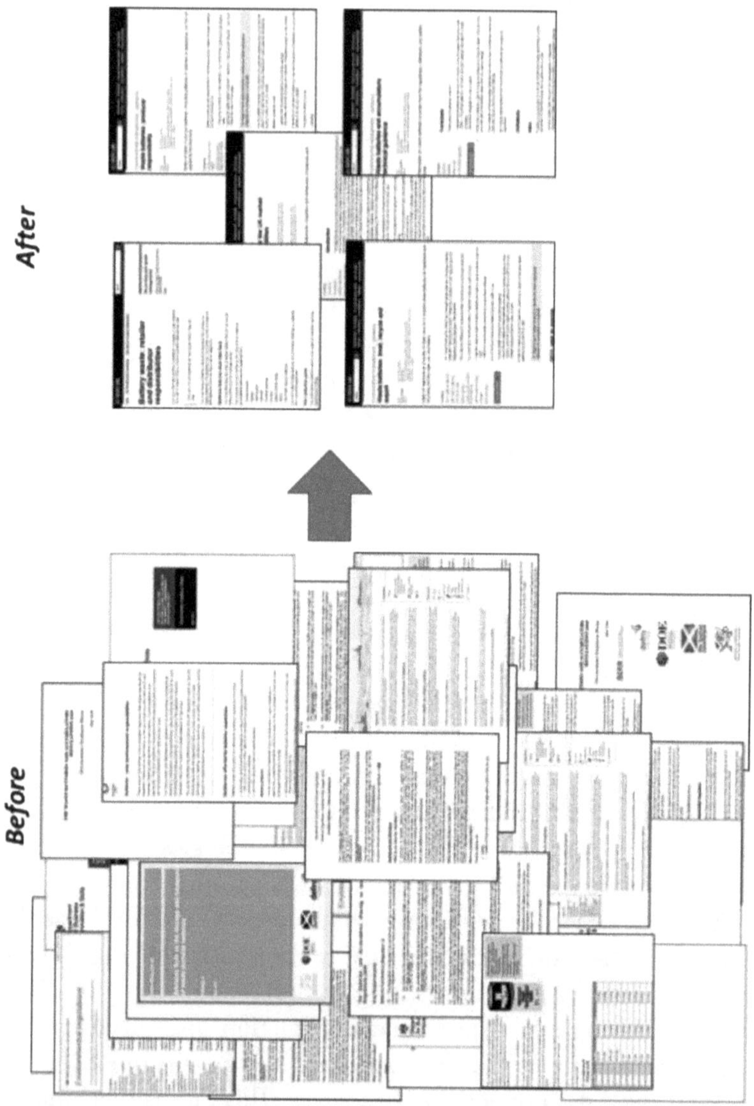

Fig. 4.3 Example of reformed guidance for batteries waste (Lockhart-Mummery, Guidance reforms in England 2016)

That is the story for one topic. The guidance reforms followed the same process for around 120 topics. By May 2015, the programme had reduced the volume of guidance by over 80%.

Embedding the Reforms

Securing the benefits of guidance reforms depends on successfully putting in place the right measures and mechanisms following the main phase of reform. Government organisations often fail to embed reforms even when they otherwise complete them successfully. Measures required from Defra to embed the reforms include[3]:

- *Complete the 120-topic-level reform plans.* While the vast majority of content has now been reformed there's still work to do to finish rewriting content around users and to archive old content. Completing this task would reduce volume by a further 50% to reach a 90% reduction in total.
- *Continuously improve content based on user performance data.* One of the benefits of digitally enabled guidance is that it is possible to quickly adjust content where it does not give people what they need. Problems can quickly be spotted through web data, such as completion rates, through the comments left at the 'Is anything wrong with this page?' facility on each page and through more detailed user testing in more complex areas. Processes need to be put in place to make sure this happens routinely.
- *Get 'browse' working smoothly.* The content reforms focussed initially on reforming individual items of content that help people complete very specific tasks—such as get a species licence. In reality people are trying to do something different, like develop a site for housing. They may not know they need a species licence and may need to do several tasks to comply with government's requirements. This makes it important to organise content so that people can easily browse and find the right content for them. Ultimately:
- *Provide the right training and staff to manage content efficiently.* This requires a small team of expert content designers and continued training for subject matter experts to help make sure content is accurate.

- *Put in place content governance and Internal processes.* This includes making sure the internal mechanisms, processes and responsibilities are in place for maintaining and improving content, and changing it when the rules change.

Beyond these essential measures to embed the reforms, there are also a number of opportunities from having user-based content. For example:

- developing 'smart tools' that ask what you're doing and give personalised content
- better ability to design digital transformation of transactional services (permits, etc.) and processes around users
- designing policy with real user experience in mind
- targeting customer contact support where bespoke help is genuinely needed

Concerns About Guidance Reform

The reforms had responded to public and stakeholder feedback, and most welcomed the reforms. However, both the principles and the scale and pace of reform met with opposition from some quarters. In addition to the debate about good practice guidance above, other points of discussion were:

- *Guidance cannot possibly be so short.* The principle is to provide what the user needs to know and no more. Sometimes that is short; sometimes, users need detail. In some cases, decisions may need to be made about whether government should publish detailed information that only serves very few users, or whether that information should be provided through other channels.
- *Regulators sometimes rely on guidance to give them a mandate to act where not covered by law.* Where this is necessary, it can be provided as a 'policy statement' so it does not clutter guidance for regulated users but can nevertheless be referred to if needed.
- *A concern that courts will not know what guidance was in force at the time of an incident.* England's GOV.UK platform allows recall of the

version that was online at the time. However, clear governance remains essential to manage and communicate changes to guidance content.

- *A view guidance should contain "waivers" in case regulators still need to enforce, despite following the guidance.* Ministers in England were strongly opposed to waivers on the basis that it is unfair and demotivating for businesses to do what government asks for and then still face enforcement action. Instead they wanted the opposite statement: if you follow this guidance, we won't enforce. It was agreed not to have any statements on the basis that reformed user-based guidance was now clearer and focussed on what's required, and it should therefore be taken as read that if you comply with it you will not face enforcement action.

4.9 England: Smarter Data Reforms

From the summer of 2013, the smarter data project reviewed all the information that government required from business or the public as part of environmental regulations and made proposals to collect the right information in the simplest possible way. This included the information collected by Defra, DECC, the Environment Agency, the Forestry Commission, Local Authorities (on Defra's behalf), Natural England, the Animal and Plant Health Agency and the Marine Management Organisation. In March 2014, the ministers announced reform measures to collect the information needed while reducing the costs to business by £30 m per annum (20% of the costs) and the costs to government by around £10 m per annum.

Why Reform Information Obligations?

Most environmental regulations involve collecting information, for example, for: permit applications, transfers, variations and surrenders; exemptions and notifications; registrations and certifications (i.e. scheme/activity); operational monitoring returns; environmental reports and assessments; and agri-environment scheme applications and claims. As regulations had developed layer by layer, so too the information they required. The Smarter Environmental Regulation Review found that:

- there was no common or co-ordinated approach to information requests either across Defra and its agencies or sometimes even within each organisation with individual teams making their own decisions about whether, how and when to collect information;
- there was no consistent approach to collecting information, with a mix of online forms, email submission, paper submission by post, indirectly via compliance schemes and portal submission;
- there were over 1000 separate information requests across the Defra and its agencies;
- these were managed on a complex web of expensive legacy IT systems;
- IT collection systems had rarely taken full account of user experience or how requests align with businesses' own collection systems;
- there was a high degree of duplication in reporting basic identification information with less duplication in the required environmental information;
- there was limited interoperability between data IT systems and limited data sharing across and sometimes within organisations; and
- it was rarely clear to businesses why information is collected and what is done with it.

Principles for Reform

An early challenge was to work out some consistent principles for what information regulators need or should collect, to reverse the historic problem, where individuals had made unbounded decisions about what information to collect and large quantities of information was never used. The principles were that information should only be collected when it is:

- essential to meeting minimum EU obligations;
- essential to making a decision/approval; or
- required to verify compliance with regulation.

and:

- No alternative and better means to achieve these are available.

- Only the minimum amount of information required to achieve the outcome is requested.
- The information collection is undertaken in the most streamlined manner available.

The essential point was that regulators should only routinely be asking for and holding information from businesses where they are actively using that information and that the means of supplying that information should be as simple as possible. This also recognised that there were situations where information might be needed, but there was no need for regulators to routinely collect it, for example, where:

- businesses need information but it does not directly inform a regulator decision;
- regulators can access information from other sources without routinely collecting it;
- the information was only needed rarely and it could be collected on demand.

Method to Overcome Obstacles to Reform

The first inherent reform obstacle was that the specialists who had developed the arrangements weren't necessarily best placed to reform them. At the same time, because the nature of environmental regulation is sometimes highly technical, complete outsiders wouldn't have the depth of knowledge to credibly challenge the status quo. Furthermore, the nature of government departments and agencies is such that the staff required to advise and agree reform measures are often so spread (e.g. from policy, legal, technical advice, IT and regulatory service teams) that it is often hard to bring them together and resolve all their practical concerns in any realistically efficient or proportionate way. HM Treasury had demonstrated this problem shortly before in running an exercise looking to reduce information obligations across the government and had found no opportunities for environmental regulation.

Defra and independent environmental consultants from WSP, therefore, developed a method to overcome these obstacles and incentivise reform as follows:

- WSP worked with technical specialists to develop initial proposals in line with the principles.
- Sector workshops reviewed the opportunities to go further to make it simpler from their perspective.
- These proposals were then scrutinised in challenge sessions where the regulatory heads of business challenge the ambition of their teams and to help work out how to overcome any practical barriers to reform.
- Proposals were put to online consultation (Department for Environment, Food and Rural Affairs 2017a).
- Proposals were then put to Ministers for agreement, except for around 20% where assessed as disproportionate.

All information obligations were divided into 'pods' and went through this reform process in sequence so that all areas were complete within nine months.

Reform Measures

There were two main types of reform measure. First, reforms to collect less or better information, including to:

- stop collecting information we no longer need or use (e.g. the Resource Efficiency Pollution Index which is no longer used for its intended purpose);
- stop collecting information we already have from other sources (e.g. providing an option to register as a waste carrier when applying for a waste exemption);
- find a more effective way to achieve the policy outcome (e.g. by more actively encouraging businesses to find responsible solutions to dispose of polychlorinated biphenyls, PCBs)
- make automatic renewals easier (e.g. for waste exemptions)

- replace bespoke licence applications with standard permits (e.g. for species licensing and environmental permitting)
- focus assessments on required information (e.g. flood risk assessments)
- standardise reporting (e.g. for environmental permit reports)

Second, reforms to simplify how data are collected, managed and used. The longer-term vision was to provide digital services that are so easy that people prefer to use them, that they never have to provide the same piece of information twice and that there is coherent information management. Features of this should include:

- information is submitted or collected from businesses in the most efficient way taking into account both business and government costs over short, medium and long terms;
- businesses can access services in one place on the web;
- businesses never have to submit the same information more than once, and information is available to those who need it irrespective of organisational boundaries;
- information is linked, and systems interoperable across regulators and the underlying data standards and architecture support this;
- coherent information can easily be made transparent.

Rather than to develop a single IT system, however, it quickly became clear working with the Government Digital Service, that the best way to achieve this was to progressively develop new digital services using a common approach in terms of consistent user, data and technology standards. The implementation plan covered the 14 largest volume environmental information obligations.

Implementation

The implementation plan was shared between Defra and agencies. The Regulated Customer Digital Programme was set up in the Environment Agency to implement the digital reforms, developing the first digital by default services and establishing digital ways of working in Defra organisations. By December 2015, agencies had implemented only around half the targeted business savings for environmental measures. The implementation plan showed that the main dependency was funding for digital projects with less than £2 m funding made available. However, Defra was awarded £66 m for digital projects in the 2015 spending review securing the resources needed to fully transform environmental data transactions during the period 2016–2020.

4.10 England: Smarter Environmental Legislation

This 2015 project co-designed a modern policy and legal framework for better integrating environmental objectives in the economy. It envisaged a stable, outcome-focussed and integrated set of arrangements to meet the environmental and social challenges of the next decades, designed for modern business realities. Given that current arrangements in England are largely prescribed in 800+ pieces of primary and secondary legislation, legislative reform seemed inevitable and the project proposed a central environment act.

Why Reform the Legislative and Policy Framework?

The terms of reference for the Smarter Environmental Legislation project explain its rationale (Department for the Environment, Food and Rural Affairs 2014c):

> Over recent decades, successive governments have built a framework of environmental legislation which has transformed the way we treat and value our environment in England. However, this framework has evolved in a piecemeal way, often in response to single issues, and it now consists of hundreds of sometimes overlapping legal instruments. Evidence shows that this framework can be fragmented, overlapping, inconsistent and complex from the perspective of those who need to comply, eroding the efficiency and effectiveness of reaching desired outcomes. This raises questions over whether a new approach could provide a step change improvement.

> Furthermore, many regulations were introduced at a time when the general assumption was that people did not consider or manage the environment responsibly. However, awareness and social norms of behaviour towards the environment have shifted so there is now far higher awareness, motivation and appetite for rising levels of environmental performance. At the same time there is greater need to make sure constrained public resources are properly targeted on genuine risks and where willful damage to the environment is caused.

Improving implementation alone would not be sufficient to make environmental regulations fit for purpose: many of the shortcomings listed earlier in this chapter are rooted in the design and structure of the legislation itself. Initially thought was given to doing an exercise to consolidate all, or at least some, of the 150 environment Acts that had been identified. While this would simplify and achieve consistency, it was argued that while absorbing large-scale resources, this would largely 'reorganise the deckchairs on the titanic'. Instead it was decided that it would be better to start by working out how to fundamentally improve the basis for achieving outcomes for the decades ahead.

The terms of reference and subsequently the recommendations contained a 'double lock' that reforms must

a. maintain and improve environmental protection
b. reduce regulatory burdens and enable businesses to do the right thing

Co-design Method

The purpose of co-designing the new framework was to benefit from the ideas and expertise from all quarters. The project was guided and championed by a high-level panel of business, NGO, government and academic leaders to secure engagement from their communities. Over 30 workshops were held starting with defining the problems and opportunities, then developing principles for a new framework, and then developing as well as testing the framework.

The project also drew on an evidence base developed in Defra between 2010 and 2014 which had reviewed international experience of environmental regulation and reform and mapped the legislative landscape in detail.

Guiding Principles for the Design of a New Framework

The principles developed for the future framework are:

- *Simple*—designed around users so people can easily understand what they need to do
- *Do once*—so people can respond easily and coherently to a single joined up system
- *Enabling*—the majority to do the right thing easily and transparently
- *Evidence based*—so it responds to real issues and is targeted
- *Co-designed*—so the best feasible solutions are found and owned
- *Predictable and outcome-focussed*—so businesses and others can innovate, plan and invest
- *Strategic*—so it targets problems at source rather than symptoms, and at the people who can and should provide solutions
- *Fair*—so it ensures a level playing field, apportions action within and across sectors in an objective and consistent way and can be enforced robustly
- *Proportionate*—so effort is focussed on what matters

The Proposed Framework

The framework was then developed through co-design sessions with the external panel and other groups working from first principles about what those who influence the shape of the environment need to be able to integrate environmental objectives into their decisions. The core components of the proposed framework are:

- *clear, stable, long-term objectives* that enable innovation and investment in solutions and infrastructure
- *an integrated approach to data and assessment* to inform a holistic understanding of issues and progress
- *long-term strategic plans to integrate environmental objectives into the economy* with environmental objectives embedded in cross government policies and industrial strategy sector plans
- *an independent body* to provide scientific and evidence-based advice for policymaking

- *integrated local planning* enabled by a single local environmental data platform
- *transformed regulation* based on a single approach to authorisation, inspection and enforcement

From Concept to Implementation

The report set out some potential scenarios and paths to implementation. Its recommended approach was to introduce an Act to establish the framework in law and then progressively reform secondary legislation and administrative arrangements. However, two unexpected events occurred in the following 15 months. First, the Conservatives won the general election in May 2015, but with a narrow majority in Parliament so any legislative reform was off the table. This project was therefore shelved. Second, the British public voted to leave the EU in June 2016. This creates the need for a domestic framework for environmental policy and legislation and has subsequently stoked interest in this project.

4.11 Scotland: SEPA's 'One Planet Prosperity' Strategy

The Scottish Environment Protection Agency's (SEPA) 'One Planet Prosperity—Our Regulatory Strategy' (Scottish Environmental Protection Agency 2016) establishes a direction and plan to fundamentally change the way Scotland regulates its environment.

The Context in Scotland

The broad global backdrop for reform is the same as for England. However, there are some notable contextual differences in Scotland. First, the pre-existing scale and complexity of regulations, and administrative arrangements is more manageable, largely as a function of a historically smaller and more coherent administration. Second, and possibly as a consequence of the first, the Scottish Government has not seen environmental regulation as in conflict with economic growth, but consistently that environmental management is essential to Scotland's prosperity. Third, the Scottish Government has therefore more consistently supported long-term sustained reforms rather than immediate reductions in red tape. Fourth, the relative simplicity of administrative arrangements makes it easier to 'get things done'. Fifth, the Scottish spirit of innovation and invention enhances the appetite within organisations for pioneering change.

The Regulatory Reform (Scotland) Act 2014 (Scottish Government 2017) sets the scene for reform. Rather than fundamentally changing environmental legislation itself, it removes legislative barriers and provides a platform for reform by:

- Giving the Scottish Government wide-ranging powers to reform environmental regulations while protecting and improving the environment. These have already been used to introduce new enforcement powers and to develop proposals for an integrated authorisation framework.
- Giving SEPA a new statutory purpose to make clear its role in serving the people of Scotland. This purpose is to protect and improve the environment (environmental success) in ways that, as far as possible, create health and well-being benefits (social success); and sustainable economic growth (economic success).

SEPA then took the initiative to publish its one planet prosperity strategy setting out how it would use the opportunity of the new Act to remake regulation fit for the challenges and context of the twenty-first century.

Defining Features for 360° Change

SEPA's CEO, Terry A'Hearn, refers to the strategy and reforms as 'a complete change in the DNA of how we regulate'.[4] Some starting points, observed by the author, that SEPA is using to define the shape and direction of reform are:

- *Change must focus on better achieving SEPA's core environmental purpose.* According to the ecological footprint measure, Scotland needs approximately three planets to sustain its current living. Businesses will not survive, and Scotland's economy flourish without finding ways of living within planetary resource constraints.

- *SEPA must make it simpler, cheaper and quicker to comply than to fail.* Non-compliance increases where it is easier to get away with avoiding compliance, than to comply. SEPA is therefore using this as a simple effectiveness test of its reforms. This means coming down hard where companies fail.
- *SEPA will enable regulated businesses go beyond compliance standards* by helping them reduce water use, carbon-based energy use, materials use and all forms of waste and pollution in ways that improve their profitability and long-term viability.
- *SEPA will only succeed if it redesigns itself around the sectors and businesses it regulates.* Whereas once government laws and regulators were the main influence, environmental performance is now driven in more complex ways by customers, employees, investors, local communities and the wider public, supply chains, trade association membership, and cost and investment considerations. SEPA will have maximum impact not only in securing compliance and beyond but on business innovation and success if it understands and works with these other influences.
- *A shift is needed from managing processes to managing outcomes.* In the world where regulation was the only influence and environmental management capabilities were low, Environment Protection Agencies developed (and focussed activities on managing) rigid processes to ensure that requirements and therefore outcomes were met. As capabilities and influences increase, SEPA will secure the best outcomes if it focuses its activities on ensuring outcomes are met.
- *SEPA will be a world-class organisation.* 'World class' is a useful yardstick both to establish the level of ambition and to inspire staff to participate in making it happen.
- *'The strategy is delivery'.* A commonly held view amongst environmental regulators is that for years we've been talking about reform but done little. SEPA staff now report: 'this time it feels different'. The strategy sets out the first set of actions. Expect more and quickly.

The strategy also contains six organisational characteristics to give all involved a clearer sense of what the future will be, and to help make it happen. They are:

1. Producing information and evidence that people use to make decisions.
2. Helping people implement successful innovation, not minor improvements on 'business-as-usual'.
3. Helping communities see the environment as an opportunity to create social and economic success.
4. Routinely interacting with regulated businesses through their boardrooms and executive teams and owners.
5. An organisation that people are clamouring to work for.
6. Using partnerships as our principal way of delivering outcomes.

SEPA's Regulatory Model

Fig. 4.4 shows the main areas where action is being taken. Sector plans are fundamental to rewiring how SEPA works, making sure that compliance and beyond compliance performance within each sector is raised based on a solid understanding of each sector (Fig. 4.4).

4.12 What Next?

This chapter sets out some early steps towards a radically improved basis for achieving environmental outcomes. Unexpectedly, Brexit now creates the opportunity, as gaps arise from our departure from the EU, to reinvent environmental policy, regulation and law so that the environment becomes a central plank and enabler for UK prosperity in terms of high quality of life and a productive competitive economy.

However, to achieve this paradigm shift will require people to participate and come together in new ways, including for:

- NGOs and business to develop a shared vision for environmental policy.
- politicians to work across party divides to develop sustained consensus
- government departments and agencies to be far bolder in driving sustained change.

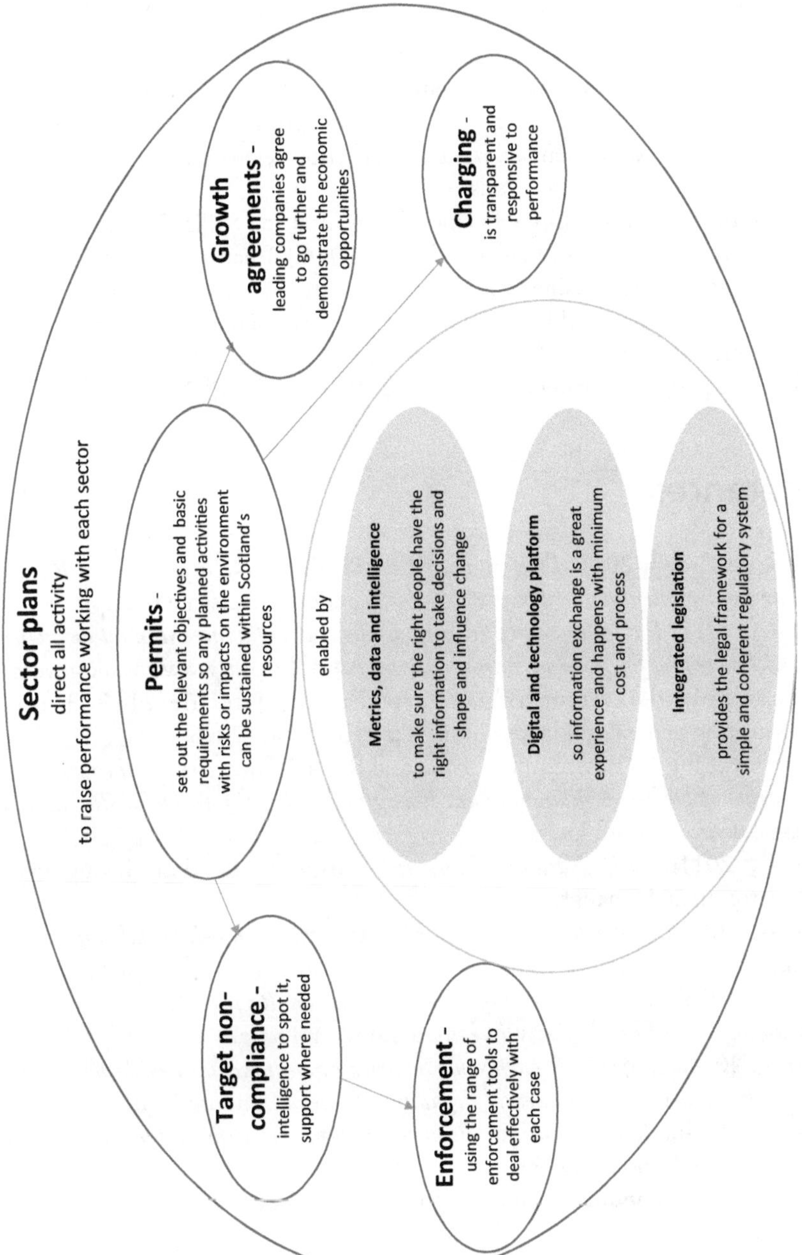

Fig. 4.4 Author's view of how SEPA's future regulatory model might work

Notes

1. Twenty-thousand pages of environmental guidance (reduced from 100,000 through the smarter guidance reforms), 200 permissions or reporting requirements, dozens of inspection regimes and 2377 environmental offences or sanctions.
2. At the time of publishing Business Innovation and Skills (BIS) and department of Energy and Climate Change (DECC) had merged to form the department for business, energy and industrial strategy (BEIS).
3. This is an expanded list from the list included by the author here https://quarterly.blog.gov.uk/2015/09/10/clearing-the-thicket/.
4. For example, meeting in Stirling at SEPA on 18 April 2017.

References

Aldersgate Group. 2011. *Dealing with Deficits: Best Value Regulation to Reduce Our Environmental and Financial Debts*. London: Aldersgate Group.

DEFRA. 2012. *Environmental Permitting. General Guidance Manual on Policy and Procedures for A2 and B Installations*. April. https://www.gov.uk/government/uploads/system/uploads/attachment_data/file/211863/env-permitting-general-guidance-a.pdf. Accessed 14 Nov 2016.

Department for the Environment, Food and Rural Affairs. 2011a. *The Costs and Benefits of Defra's Regulatory Stock. Emerging Findings from Defra's Regulation Assessment*. London: Defra.

———. 2011b. *UK National Ecosystem Assessment*. London: Living with Environmental Change.

———. 2013a. *Smarter Guidance Method Statement—Internal Operating Instructions*. London: DEFRA.

———. 2013b. *Smarter Environmental Regulation Review: Phase 1 Report—Guidance and Information Obligations*. London: Defra.

———. 2014a. *Terms of Reference for Smarter Environmental Legislation: Phase 2 of the Smarter Environmental Regulation Review*. London: Defra.

———. 2014b. *Defra Better for Business. A Strategic Reform Plan for Defra's Regulations*. London: Defra.

———. 2014c. London. Legislation.gov.uk

———. 2017a. *Homepage*, May 4. Accessed on Smarter Guidance and Data Consultation. http://webarchive.nationalarchives.gov.uk/20150423095755/http:/guidanceanddata.defra.gov.uk/

———. 2017b. *Smarter Environmental Legislation (Final Report)*. London: Defra.

EU Member States. 2017. *Homepage*. May 5 Accessed on Make It Work Project. http://www.ieep.eu/work-areas/environmental-governance/2014/11/make-it-work-project-for-smarter-eu-environmental-law

European Commission. 2013. *Decision No 1386/2013/EU of the European Parliament and of the Council of 20 November 2013 on a General Union Environment Action Programme to 2020 'Living Well, Within the Limits of Our Planet*. Brussels: European Commission.

European Environment Agency. 2015. *State of the Environment Report. State and Outlook. Global Megatrends*. Copenhagen: European Environment Agency.

Global Footrpint Network. 2017. *Country Metrics*. May 4. Accessed on Global Footprint Network. http://data.footprintnetwork.org/countryMetrics.html

HM Government. 2010. *Reducing Regulation Made Simple: Less Regulation, Better Regulation and Regulation as a Last Resort*. London: HM Government.

———. 2011. *The Natural Choice: Securing the Value of Nature (The Natural Environment White Paper)*. London: HM Government.

———. 2017. *Homepage*. May 4. Accessed on Red Tape Challenge. http://webarchive.nationalarchives.gov.uk/20150522175321/http:/www.red-tapechallenge.cabinetoffice.gov.uk/home/index/

HM Treasury. 2011. *The Plan for Growth*. London: HM Treasury.

Lawton, S.J. 2010. *Making Space for Nature: A Review of England's Wildlife Sites*. London: Defra.

Lockhart-Mummery, E. 2015. *Clearing the Thicket*. September 10. Accessed on Civil Service Quarterly. http://quarterly.blog.gov.uk/2015/09/10/clearing-the-thicket/

———. 2016. *Guidance Reforms in England*. London: Resource Decisions Ltd.

Nielsen Norman Group. 2006. *F-Shaped Pattern for Reading Web Content*. April 17. https://www.nngroup.com/articles/f-shaped-pattern-reading-web-content/

Scottish Environmental Protection Agency. 2016. *One Planet Prosperity—Our Regulatory Strategy*. Stirling: SEPA.

Scottish Government. 2017. *Regulatory Reform (Scotland) Act 2014*. May 4. Accessed on Legislation.gov.uk. http://www.legislation.gov.uk/asp/2014/3/pdfs/asp_20140003_en.pdf

The National Archives. 2017. *Defra Search Results (Environment and Marine)*. May 4. http://www.legislation.gov.uk/defralex
UK Environmental Law Association. 2012. *The State of Environmental Law in 2011–2012*. London: UKELA.

Edward Lockhart-Mummery is an environmental economist by background who has spent the last 15 years developing economic assessments and leading reform initiatives across the spectrum of environmental policies for UK and EU governments and agencies. He has developed a number of cross-cutting initiatives including the assessment of all the UK Department for Environment Food and Rural Affairs (Defra) regulations. He led the Smarter Environmental Regulation Review for Defra and agencies and implemented its recommendations including smarter guidance, smarter data and smarter environmental legislation. He also led Defra and its agencies' initial work on digital transformation. He is advising several European governments and agencies on environmental policy and reform.

5

The Sector-Based Approach and Partnerships: Regulatory Interventions to Reduce Risk and Promote Compliance

Jon Foreman

5.1 Introduction

This chapter takes a detailed look at the development of the sector-based approach used by regulators and shows how this has evolved to become a key part of the way regulators organise themselves to engage with industry. The chapter looks at the rationale for the sector approach, how sector thinking is integrated with other related activities and the pros and cons of the approach. The chapter illustrates a range of stakeholder-based sector initiatives and programmes which when taken together have had a significant influence on business and led to the improvement of environmental management and the delivery of regulatory outcomes. Examples of regulator initiatives are given along with a discussion of the way the Environment Agency (EA) in England organises its regulatory work along industry sector lines.

J. Foreman (✉)
Jon Foreman Associates, Bristol, UK

J. Foreman (ed.), *Developments in Environmental Regulation*, Palgrave Studies in Environmental Policy and Regulation, DOI 10.1007/978-3-319-61937-8_5

5.2 The Evolution of Sector-Based Thinking in Environmental Regulation

The development of sector-based thinking by policymakers and regulators can be traced back 15–20 years or more and reflects the introduction of formal 'integrated' regulatory regimes that target certain industry sectors. In the UK, the introduction of Integrated Pollution Control (IPC) from 1990 and then the EU-wide Integrated Pollution Prevention & Control (IPPC) and Industrial Emissions (IE) Directives each set out on a sector basis the 'Prescribed Activities' that were subject to regulation.

For some sectors, production or capacity thresholds defined whether a process or activity would be subject to regulation. The most potentially polluting 'process industries' were subject to the full requirements of the legislation, whereas other sectors were either only partially regulated or left entirely outside the scope of direct regulation. Earlier chapters have highlighted the broad sectors subject to environmental permitting requirements. Industries such as general construction, engineering and agriculture were typically outside of the scope of regulation. Such 'non-regulated' sectors would often only come to the attention of regulators when environmental incidents or the scale of numerous smaller activities demonstrated that the sector as a whole had a significant environmental impact.

As early as 1995 in the USA the national administration set out a vision for future environmental regulation that put collaborative sector initiatives at the heart of its thinking. The paper 'Reinventing Environmental Regulation' by President Bill Clinton and Vice President Al Gore (USEPA, March 1995) describes, among other initiatives, the setting up of common sense 'Strategies for Sectors'—industry covenants to demonstrate how adjustments in environmental regulation can achieve cost-effective results for the environment. Other principles outlined in the paper which chime with the sector approach included the development of strong and trusting relationships, the use of information, regulator and industry accountability and responsibility. These are key themes which regulators have developed in more recent demonstrations of the sector approach.

The EA for England, and to a large extent the other UK and EU regulators, take a 'sector based approach' to regulation, dividing regulated industries into broad sectors, where the commonalities between the businesses in the sector and the way they are regulated allow for a national coordinated approach. There are of course many ways by which sectors can be organised, but by and large, this has been done by following or grouping related 'Activities' as listed under the relevant legislation, in particular, the Industrial Emissions Directive (IED).

Notwithstanding the steer provided by the legislation, the parameters that regulators have considered in the development of sector initiatives are varied and may be subject to some debate. Some industries, for example, may have more than one prescribed activity, say a chemical process and a coating process which would be listed under different sectors in the legislation. Where this occurs, usually the primary activity takes precedence in determining which sector a site is assigned to. While often defendable, prioritisation criteria may reflect political and organisational factors as well as objective evidence-based realities in determining which sectors to focus on. Some key considerations for sector selection are shown in Table 5.1.

Table 5.1 Criteria for sector-based initiatives

Scale	The size, significance, number and distribution of the sector. Economic measures
Variability	Range of activities and process types within a particular sector
History and direction	Is it a growing or declining sector
Environmental significance	Emissions, waste and effluent volumes, pollution potential
Governance and politics	How well managed is the sector. Will the industry cooperate with regulators at the strategic and local level?
Representation	Does the 'sector' have a coherent trade body or organisation(s) that can effectively represent the interests of all players in the sector without causing conflict
Policy and legislation	The extent and development of regulation, and whether new or proposed regulations affect the sector
Regulatory compliance	How compliant is the sector? Have there been a significant number of prosecutions or environmental incidents

5.3 Stakeholder and Sector-Based Best Practice Programmes and the Link with Regulation

Prior to the development of formal regulator-led sector-based approaches, which in the UK developed from around 2010, regulators and governmental bodies often engaged with industry sectors on a collaborative basis, creating formal or informal partnerships to further environmental and business objectives.

In these projects, the regulators would often be able to promote regulatory compliance through information sharing and the provision of advice and guidance, but typically, there would be no mechanism for ensuring that the individual companies in the programme complied with the legislation. While the aim of these initiatives was often on waste reduction and resource efficiency, the industry focus provided significant insight and experience for both regulators and regulated business. Indeed, many initiatives were co-funded or supported by regulators such as the EA or the Scottish Environmental Protection Agency (SEPA) as well as Local Authorities. This gave regulators direct access to groups of regulated businesses where topical information on regulation could be imparted in an informal non-regulatory environment.

For some initiatives, barriers were put in place to ensure the regulators were not given access to confidential company information which would enable them to check compliance. In other instances, the regulators were invited into the heart of the project to ensure that regulatory considerations were identified and dealt with before they became a barrier to development. In 2005, the UK National Industrial Symbiosis Programme (NISP) programme (www.nispnetwork.com) received government funding and endorsement to employ a senior EA advisor to manage inter alia regulatory issues that arose particularly over the definition of waste—an issue which is often cited as a barrier to the re-use and recovery of industrial wastes and by-products.

In the decade prior to the financial crisis, large governmental programmes and collaborative initiatives were commonplace with ample funding available through the EU, government, and regional funds as well as from sources such as the Landfill Tax credit scheme (ENTRUST Landfill Communities Fund Statistics). This led to an explosion of environmental good practice initiatives, many of them targeted at industry sectors.

Waste Minimisation/Resource Efficiency Programmes

From around 1994, the UK government funded several national and regional programmes aimed at improving business efficiency and reducing waste. The Environmental Technology Best Practice Programme (ETBPP), which later became the 'Envirowise' programme, pioneered 'demonstration' 'Waste Minimisation Clubs' which targeted industry either geographically (such as the West Midlands) or at an industry sector such as the food and drink sector. At the height of their popularity, there were over 60 active waste minimisation or Resource Efficiency Clubs (RECs) active across the UK with each supporting a network of member companies (Phillips et al., August 1999). A range of different engagement models were adopted by these initiatives with a number focussed on either wholly or largely on industry sectors (Defra, November 2011).

In addition to the REC's programme, the ETBPP/Envirowise provided a wealth of best practice guidance and tools, much of it targeted at sectors industries, and often where there was a coincident driver for cleaner production. The introduction of new regulations often provided this demand and an opportunity for regulators to not only further best practice but to promote regulatory compliance. For example, the imminent introduction of new hazardous waste regulations and IPPC formed the basis for a joint initiative between the Metal Finishing Industry, consultants, and the EA to initiate an 'IPPC club'. The club provided expert-facilitated workshops for industry clients, where the key aspects and information requirements for permit applications were explained (HANSARD, February 2004). Such initiatives have helped to secure industry engagement and political support for what can be challenging circumstances, especially for smaller businesses.

Internet, Sectors and Regulation: NetRegs

In the late 1990s, the UK environmental regulators developed an extensive internet-based regulatory information hub known as NetRegs which was targeted at small and medium-sized enterprises (SMEs). Information was presented on a sector basis and the project ran for over 10 years across the UK until funding cuts in England curtailed it as a UK-wide project. NetRegs is still funded and available in Scotland and Northern Ireland as a partnership between the Northern Ireland EA (NIEA) and Scottish Environment Protection Agency (SEPA) and provides information and tools to a range of SME-dominated business sectors. NetRegs can be found at: www.netregs.org.uk (Box 5.1).

Box 5.1 NetRegs vision

NetRegs is a free-to-use website which aims to help SMEs in the UK to understand the complex environmental regulations that can affect them. It is aimed at any SME operator that might need to comply with any environmental regulatory requirement. The site provides guidance on how to comply with environmental law as well as advice on good environmental practice. It provides clear, readily accessible information to businesses on the environmental legislation that affects them. This helps a business take the first step towards environmental compliance and resource efficiency.

The site comprises four main areas:

- Sector-specific guidelines for over 100 sectors;
- Management Guidelines covering different aspects of business operation;
- Current legislation: detailing regulations in all regions of the UK;
- Future legislation: including consultations and EU law developments.

Source: Netregs Case Study 4, European Commission (2006)

NetRegs became recognised as a reliable and extensive source of information on regulation and best practice guidance. But as a programme, NetRegs was separate from the formal regulatory policy and compliance functions of the regulatory agencies. This led to some duplication with the regulators' main websites. For this and other reasons, including a perception of information overload, the UK Coalition Government (2010–2015) set about a rationalisation of internet-based environmental information. A new governmental website was established and both NetRegs and the

EA (for England) had their extensive web-based guidance information abolished. Much of the essential regulatory guidance was retained but with less contextual information and with the loss of a great deal of best practice information and Codes of Practice. This was generally regarded as a retrograde step by regulators, advisors and industry representatives alike. Now, all official UK regulatory information can be found by searching relevant criteria on the government website: www.gov.uk.

Hazardous Waste Sector Initiative HAZRED

Changes to the EU Hazardous Waste Directive and associated regulations in member states in 2005 led to the EA working with Ireland's Environmental Protection Agency and others to establish a project to help regulated businesses reduce priority hazardous wastes. See Box 5.2 for details.

Box 5.2 HAZRED—Hazardous Waste Reduction for SMEs: A Case Study of Sector Prioritisation

HazRed was an EU-Life-funded partnership project led by the EA aimed to demonstrate the benefits of a sector-led approach to working with small businesses to reduce hazardous waste.

New Hazardous Waste Regulations which came into force, in July 2005, expanded the list of substances classed as hazardous. Thus, many small businesses were classed as hazardous waste producers and faced increased regulatory and disposal costs. The project sought to develop a methodology to identify those SME business sectors most affected by the regulations, and where the potential for hazardous waste reduction was the greatest and to then provide support in identifying and implementing ways to manage and prevent hazardous waste.

HazRed worked with 120 businesses across six priority sectors to set hazardous waste reduction targets and develop waste reduction plans. The six sectors chosen following a prioritisation methodology were:

- construction;
- treatment and coating of metals;
- maintenance and repair of motor vehicles;
- photographic processing;
- manufacture of speciality organic chemicals;
- manufacture of machinery and equipment.

The sector selection process was done in close consultation with a wide range of environmental and industry experts. The methodology included a multi-criteria analysis scoring tool which took account of the number of small businesses in each sector and the volume and type of hazardous wastes produced and balanced that against expert opinion.

Source: Irish EPA 'Reducing Hazardous Waste—The Hazred Project'

The Food and Drink Industry

In the food and drink sector, there has been considerable effort to promote wider environmental goals through industry-wide partnership initiatives. Early work by the Envirowise programme led to the development of a wealth of technical guidance material aimed at helping manufacturing businesses across the sector to improve efficiency and reduce wastes. The Courtauld Commitment, a voluntary agreement with over 120 signatories, joins food retailers and manufactures in a sector-wide initiative to tackle key issues on waste and climate change. This resource efficiency programme, facilitated and managed by Waste and Resources Action Programme (WRAP), has evolved over the years and now as 'Courtauld 2025' is described by WRAP as an 'Ambitious, collaborative action to cut the resource needed to provide our food and drink by one-fifth over ten years' (Courtauld Commitment) (Accessed May 2017).

Courtauld 2025 now includes objectives to reduce water use, which until 2014 was the focus of the Federation House Commitment (FHC). This water efficiency initiative provided in-depth technical support and encouragement to food and drink manufacturers. As an example of co-regulation, the EA worked closely with the project to facilitate technical advice to food and drink businesses that were regulated under the IED. Where regulated businesses were party to the FHC and making progress, the EA stepped back from its regulatory scrutiny. Unfortunately, the programme was closed in 2014 because of Defra budget cuts though some information is still available through the WRAP website.

5.4 Drivers of the Sector-Based Approach

The benefits for regulators and industry of the sector-based approach can include:

- Efficiency and effectiveness of regulation—optimising regulatory costs and burdens, taking a more targeted approach to issues, developing regulator and industry expertise to solve problems
- Evidence-led improvement of environmental aspects and delivery of key outcomes, for example, reduced pollution incidents, less waste, improved compliance and reinforced by regular performance reporting. Plans provide a framework for setting goals and targets and measures for securing improvement.
- Customer focus and cooperation—responding to and reflecting industry feedback, engaging with industry to help deliver outcomes. Regulators have an improved understanding of the complexities of industry and industry better appreciates the benefits and requirements of regulation
- Improved consistency—regulators more accountable and subject to management direction
- A more holistic approach—working across disciplines rather than in established regulatory silos enabling knowledge transfer and the management of issues on a life cycle basis, for example, waste flows, pollution trends, product-related issues and climate change.
- Allows for the development of a strong partnership between the regulator and the regulated community and provides an efficient model for the delivery of environmental outcomes.

Set against these advantages, a sector-wide approach (as opposed to an exclusively site-based approach) may risk diverting regulatory resources away from activities and sites where most attention is needed. Sector organisations tend to represent the generally compliant industries and may have less influence over those businesses that flout the law than the regulator themselves. For these reasons, the sector-based approach needs to work in harmony with the site-based approach by providing the context for the targeting of regulation and by embracing the assistance of industry representatives to reinforce regulatory requirements.

5.5 The Development of Sector Plans in England

From the mid-2000s, Sector Plans for specific industry sectors were developed by the EA including the Waste Sector Plan 2006, the Cement Sector Plan 2005 and the Nuclear Sector Plan 2005. The plans and their progress reports detail the actions taken and the achievements of these plans. These early sector plans were coordinated and written by the EA and involved considerable high-level engagement with industry and set out agreed objectives, targets and actions. Inevitably, the development and maintenance of such plans was highly resource intensive, and this became a major challenge under the financially constrained environment after the financial crisis of 2008. The aims and objectives of The Sector Plan for the Waste Management Industry 2006 illustrate the thinking at the time (Box 5.3).

Box 5.3 The Purpose of Sector Plans

Focus on the most significant risks the sector poses and its impacts on the environment; improve the sector's environmental management and performance; prioritise and target regulatory effort; through cooperation, achieve environmental benefits beyond those achievable through 'conventional' regulation; monitor progress in delivering the environmental improvements (Environment Agency 2006. The Sector Plan for the Waste Management Industry 2006. Foreword, Dr. Paul Leinster, Acting Chief Executive).

Following publication of the agreed plans, industry trade organisations reported progress against the objectives and targets set out in the plans. A few progress or performance reports for these plans are available online: the Nuclear Sector Plan (Environment Agency 2012), the Waste Sector report (ESA and Environment Agency 2010) and the Cement Sector Plan (Mineral Products Association (MPA), November 2011). Cynics may argue that reports such as this offer industry the opportunity to greenwash their performance but, on the other hand, the reports reflect a significant commitment to deliver on the agreed objectives in the plan. For the cement sector, progress against the following plan objectives are highlighted in the 2011 report:

1. To increase the use of waste used as raw materials or fuel in cement works (resource efficiency)
2. To reduce waste disposal from cement manufacturing
3. To reduce air pollution from cement manufacturing
4. To reduce emissions of greenhouse gases per tonne of cement
5. To improve regulatory compliance and stakeholder perception of sites

Beyond the formal regulatory regimes of IPC or IPPC/IED and the sector-wide permitting and compliance requirements of the Environmental Permitting Regulation's (EPR), more diverse sectors such as Construction and Agriculture, have been subject to little direct regulation. Certain generic requirements such as Control of Pollution (Silage, Slurry and Agricultural Fuel Oil) Regulations and the Waste Management Duty of Care can apply to any business though these are often only passively enforced due to the lack of revenue.

Even so, these sectors have often been targeted by regulators because of the significant environmental impact they can have. Sectors may come to the attention of policymakers and regulators for a range of reasons, but where there are significant and consistent complaints, pollution incidents and prosecutions which drain regulatory resource and consistently fail to get to the 'root cause' of issues, the political and business rationale for collaborative action can be overwhelming.

Hence despite the lack of direct legislative mandate, collaborative approaches have often provided the best option for regulators as can be seen from the number of initiatives in these sectors over the years.

Similarly, in the UK, the water industry as a sector is only partly regulated under the EPR regime but is subject to a range of environmental legislation, including the Water Industries Act 1991.

Sectors Regulated by the UK Environmental Regulators

The EA and other UK environmental regulators oversee a wide range of business sectors, organisations and activities, including principally:

- Waste management—waste storage, treatment, transfer, use, landfill, incineration and bio-waste
- Manufacturing industry—including chemicals, food and drink, metals, power generation, refineries and fuels, paper and textiles, cement and minerals, construction and mining
- Radioactive substances—the use, storage and disposal of radioactive substances
- Water companies—discharges from sewage treatment and sewerage systems, and the abstraction of water
- Producers of packaging, batteries, waste electrical and electronic equipment, and vehicles
- Agriculture, discharges to surface and groundwater, water abstraction, the management and disposal of farm wastes, and spreading of waste on land, disposal of pesticides and sheep dip, intensive farming (pig and poultry) units, and the storage of silage, slurry and fuel oil.

These sectors vary significantly in size and distribution as shown in Chart 5.1.

The approach of all UK regulators is based on targeting activities and sites that pose the greatest risks, in order to secure the best outcomes while making the most efficient and effective use of resources, and reducing the administrative costs on businesses (Environment Agency 2013). The approach reflects the five guiding principles of 'good regulation' as set out by the Better Regulation task force in 2003 (Cabinet Office 2003), namely:

Proportionality—Regulators should only intervene when necessary.
Accountability—Regulators must be able to justify decisions and be subject to public scrutiny.

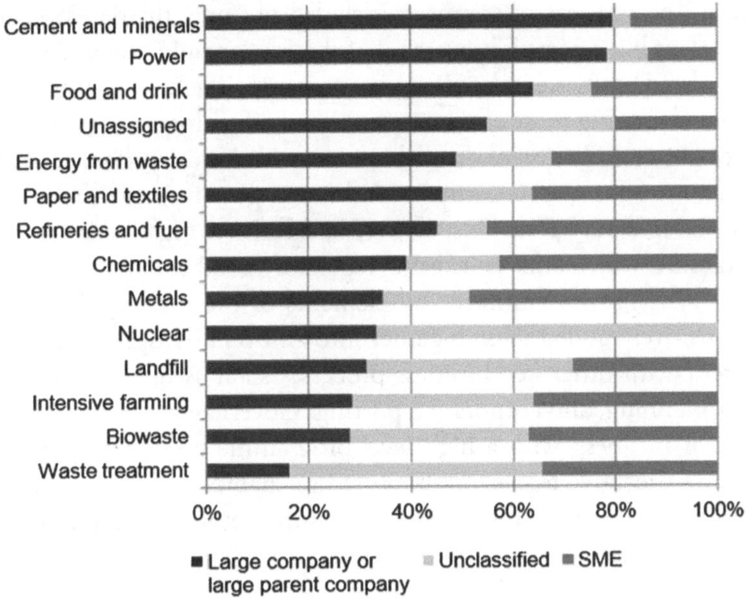

Chart 5.1 The size of regulated industries operating in different sectors. Environment Agency (2013a)

Consistency—Government rules and standards must be joined up and implemented fairly.

Transparency—Regulators should be open, and keep regulations simple and user-friendly.

Targeting—Regulation should be focussed on the problem and minimise side effects

5.6 Environment Agency: 'A Future Approach to Regulation' and the Sector-Based Approach

Following the development of early sector plans and initiatives as illustrated above, the EA (England and Wales at the time) established a programme in 2009/2010 to develop and roll out a sector-based approach under the banner of 'A Future Approach to Regulation' known in the EA as the 'FAR' project. The FAR-sector approach included not only the development of sector plans but a complete overhaul of the EA's organisation and priorities for regulated businesses. The plans were written by the EA who were responsible for delivering most of the actions. While the programme involved an element of consultation and collaboration with industry sectors, the FAR programme was more concerned with the arrangements and effectiveness of the EA's regulatory functions. It was, in effect, an organisational change programme, which built on existing regulatory good practice.

The FAR project embraced all elements of the EA's regulatory work and effectively pulled these together into an overall programme that was coherent with other key business processes such as financial and manpower planning, and corporate reporting. Governance was provided by a 'Head of Business' with a high-level programme board providing oversight and executive managers appointed to champion each sector.

The project itself ran for some 18 months before the sector approach became 'business as usual' in 2012 (Box 5.4).

Box 5.4 Key principles for the FAR project

- Staff have a high level of expertise and were equipped to challenge industry, thereby ensuring a high level of protection to the public and the environment. This could mean fewer but more skilled staff spending more time working on a particular sector
- Smarter regulation—intervening only when absolutely necessary—following the principles of good regulation outlined above and supporting industry to achieve their ambitions. This might involve using 'alternative' and complementary approaches to traditional permit and compliance-based regulation.
- A focus on customer service and empowering industry to promote innovation and growth while providing greater certainty for business through the dissemination of annual action plans
- Clarity on outcomes to be achieved and the regulatory actions and priorities that would deliver them
- Expenditure matching funding—improving efficiency and minimising cross subsidies, where the income from some lightly regulated sectors was subsidising greater regulatory attention in others.

Source: Environment Agency (2011)

The sector approach of the EA and to some extent the other UK regulators incorporates the following elements:

- Planning and coordination: strategic and annual sector action plans incorporating Specific, Measurable, Achievable, Realisable, Time-bound (SMART) objectives, targets and key performance indicators
- Evidence and intelligence led: use of latest industry performance data to determine national and regional risk-based priorities and actions; development and trialling of innovative approaches which are tailored to sector conditions
- Coordination and management: a national expert sector group, management board with regular monitoring and performance reporting (annual scorecard reports and industry challenge)
- Training and development of regulators and industry: joint working with industry, for example, on Codes of Practice and site visit to raise competency and awareness, work on specific technical capabilities for each sector
- Local management and accountability of regulation allowing for local flexibility within the nationally agreed action plan
- Coordination of technical standards and input to EU BAT reference Document (Bref) and permit reviews
- National communications: of priorities, plans and initiatives internally and externally to industry with formal trade association liaison and consultation
- Financial and manpower discipline, ensuring the right staff were allocated and that the sectors resources (income) were tracked and expenditure maintained within budget
- Added value; promoting compliance and join up with other EA regulatory and advisory roles such as water management planning, climate change adaptation, pollution prevention and energy efficiency
- Corporate reporting and links with EA strategic objectives and performance measures—such as targets to reduce the number of D-F compliance ratings and reduce the number of pollution incidents (Table 5.2).

The Environment Agency Sector Organisation

The diagram below illustrates the internal structure and connections of the relevant EA functions that together are responsible for the delivery of the sector approach (Fig. 5.1).

Table 5.2 Environment Agency FAR sectors

• Nuclear	• Water companies
• Non-nuclear	• Smaller water users
• Combustion	• Waste storage, treatment, transfer and use
• Refineries and fuel	• Biowaste treatment and use
• Metal manufacturing	• Landfill
• Cement and minerals	• Agriculture
• Chemicals	
• Paper and textiles	
• Food and Drink	

Source: Environment Agency (2011)

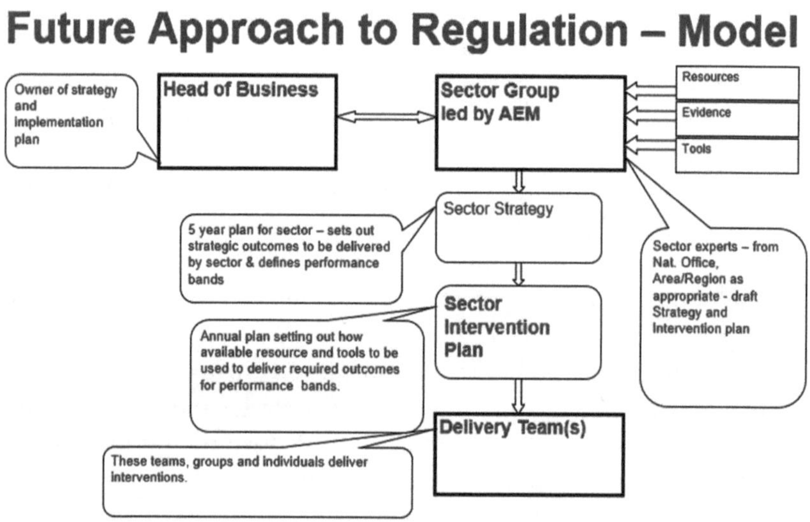

Fig. 5.1 Future approach to regulation model. Source: Environment Agency (2011)

Sector Groups—Membership, Organisation and Management

A core arrangement of the EA's sector approach has been the establishment or empowering of national 'sector groups' chaired by an Area Environment Manager and coordinated by a national sector coordinator who was also the national technical and regulatory lead for the sector. Other members of the sector group would be drawn from the inspectorate pool across the country, usually experienced officers with a detailed knowledge of the sector or aspects of it. Typically, a dozen or so officers would make up the core sector group. Other inspectors from across the country might spend part of their time focussed on a particular sector and would take their direction from the sector group. Not all EA officers would be members of a sector group, though one of the aims would be to encourage development opportunities where local staff can gain experience and exposure to national issues, thereby improving their capability.

The sector group was ultimately responsible for the planning and prioritisation of sector objectives which were directed through an annual 'action plan' which was monitored closely and subject to scorecard reporting. Other responsibilities included facilitating and monitoring staff development and training which led to some groups organising national conferences, specialist training days and site visits to support staff.

The stated terms of reference for sector groups included the following:

- Drive environmental improvement
- Promote awareness of sector issues
- Ensure consistency of regulation across the sector
- Commission and develop tools and undertake research
- Lead on best available techniques (BAT) and technical standards
- Monitor staff numbers and capabilities

Typically, sector groups would meet every two to three months for face-to-face meetings where progress against the sector plan objectives would be discussed along with wide range of relevant technical and regulatory issues pertinent to the sector. A typical agenda for a sector group meeting might include:

Sector 1—Meeting Agenda (typical)

- The plan
- Workforce planning
- Training requests
- National conference
- Roles and responsibilities
- Regional issues
- Tracking plan and recording successes

Each member of the sector group had a 'terms of reference' document that set out their respective roles and responsibilities. Key members were:

Sector chair—Senior Environment Manager
Sector coordinator/Technical specialist
Head of business (strategic policy) representative
Senior front-line inspectors and 'sector facing' officers

Sector Strategies and Plans

A five-year strategy for each sector sets out key issues to be addressed through permitting and compliance activities, detailing:

- Sector structure and economic overview
- Performance banding
- Regulatory issues (including future legislation)
- Local, national and international issues
- Link between corporate strategy and sector outcomes.
- Targets and milestones.

Annual action plans were originally described as 'intervention' plans, but that term was soon dropped as it was poorly perceived by industry.

Actions set out in the plans were to be followed through the EA permitting and compliance activities. The plans incorporated:

- Geographical distribution of sites/activities
- Compliance rating of sites/activities by region
- Overview, economic, environmental and performance for whole sector (general trends)
- Income
- Actions, resources and success measures
- Audit and inspection frequencies

In March 2017, the EA published sector summaries for 13 'Regulated Industry Sectors' (Environment Agency 2017) as part of the Annual Evidence summary for 2015. These were described as 'high level documents containing the strategic objectives of each sector for the next four years and supporting data'. An example for the Food and drink sector is shown in Appendix 1. An example of an Annual Action Plan is given in Appendix 2. Industry sector performance data for the sectors regulated by the EA can be found in the Annual Evidence summary reports.

Head Office Sector Coordination and Communications

Nationally, a 'chief fixer' role was established in the head office policy unit to provide coordination and timely updates on policy positions from the 'Head of Business'. Regular communications and updates were provided along with instructions and an annual calendar of key deadlines to ensure that expectations were clearly understood and acted upon. This head of business role also collated quarterly performance data for the management board to ensure sectors were 'on target' with their various commitments.

As a major function of the regulatory agencies, several corporate Key Performance Indicators (KPIs) related to the regulatory function. Where relevant, these were incorporated into the objectives of the Sector Plans, so, for example, sectors became committed to the delivery of national objectives. Examples include KPIs to reduce the number of

'Poor' compliance ratings (Operational risk appraisal, OPRA bands D, E and F) and to reduce the number of serious pollution incidents. These high-level KPIs help drive the actions of the sector plans, as they are subject to quarterly performance reporting. Further information on the EA's corporate reporting can be found in the regulators' annual reports (Environment Agency, Annual reports) and in relevant EA board papers (Environment Agency, governance information).

Operational Delivery

Operational delivery and day-to-day regulatory priorities is directed by local managers whose teams comprised officers who are often responsible for a range of sites from different industry sectors. Some 'sector facing officers' may spend a greater proportion of their time on particular sectors but generally speaking inspectors are able to transfer their approach to other sectors. This arrangement was a subject of some considerable debate during the Future Approach to Regulation (FAR) project, but the overall consensus was that it is better to allow officers to have exposure to different industry sectors rather than to specialise in a narrow area.

To help ensure sector plans were embedded into operational delivery, shortened 'sector summaries' were linked to the national KPIs so that key expectations were understood and minimum requirements were delivered.

Industry Consultation and Liaison

In addition to the national internal sector groups, each sector has arrangements for regular industry liaison meetings with relevant trade association representatives. These meetings will be held at a frequency relevant for the sector and would typically be a quarterly arrangement. A significant part of the agenda would be dedicated to updates on the EA's sector work with feedback and analysis on industry performance to challenge poor-performing sectors (or sub-sectors), to take forward joint initiatives and to give advance notice of EA regulatory priorities as set out in the sector action plans.

Systems, Data and Information

Annual performance reports were generated from a wide range of regulatory data-sets that are held and managed by EA, with key data-sets including:

> Compliance database; National Pollution incidents; Pollution inventory (detailing prescribed emissions and waste transfers); Enforcement and prosecution databases.
>
> To ensure accuracy in both reporting and to ensure the data and information can be used intelligently for directing regulatory activities, a considerable amount of regulator and back office time is spent on data entry and data clean-up and analysis.
>
> To support the development of sector strategies and plans, and to gain a sense of economic perspective, economic summaries of the main sectors were produced outlining the macro economic conditions and pressures affecting each sector. This information also enabled the EA to take account of political pressures and the emphasis on enabling and protecting economic growth as reflected in the Coalition Government's 'Growth duty', for example (BIS, BRDO, January 2014).

Performance Reporting

In the EA, an annual performance update or sustainable business report is produced detailing compliance and pollution trends for each sector with analysis of causes and trends. In the early 2000s, this report was known as the 'Spotlight' report and then the 'Sustainable Business' report (Environment Agency 2013). More recently, this information has been released as '"Regulating for people, the environment and growth"—2015 evidence' (Environment Agency 2016).

These reports would usually be launched with publicity to raise awareness of key regulatory issues and challenges. Such reports have been released via press release with various levels of sector information communicated directly to industry via the established trade association channels. Most recently, 'Evidence summaries' of overall regulated industry have been produced (Environment Agency 2015b).

The sector approach is not without its challenges and practitioners may raise concerns such as the diminution of the workforce, reduced flexibility for staff to work across sectors and pressure to work on 'important' topics that may not necessarily be funded by regulatory charging schemes (Table 5.3).

5.7 Related Regulatory Initiatives

There are several initiatives regulators have pursued that help improve the efficiency of the sector-based approach including, for example:

(i) *Account Management*

In parallel with the development of the sector approach, regulators have worked with some of the larger companies that hold a significant number of environmental permits by developing an 'account manager' approach. Here, high-level meetings are held between the regulator and the company, and discussions held on the detail of recent company performance as well as regulatory positions and approaches are discussed. These meetings are often fruitful in helping to move forward on persistent problems or sites where local discussions may have reached an impasse. For a number of sectors, the use of account management has been one of the key actions in helping to raise the performance bar of companies with a poor compliance track record.

(ii) *CLEAR info*

In 2011–2014, The UK EA led an EU LIFE+ project known as *Company Level Environmental Accountability Reporting and Information* (CLEAR Info) (Environment Agency 2015a) to explore the potential for using integrated compliance data to drive compliance with European environmental legislation.

The project built an innovative data collation tool to match site-level compliance data with company information. It created and used data and reports to engage with financial and investment specialists, large companies with multiple permits, European environmental regulators and entrepreneurs.

Table 5.3 The EA's sector approach: pros and cons

Advantages	Disadvantages
Coordinated national 'one voice' approach with targets and actions agreed nationally.	Tension with local office as national targets and objectives may not always reflect local issues.
Greater efficiency and effectiveness with less hobbyists where individual regulators pursue their own interests or priorities.	Local inspectors expected to follow the national line may become discouraged. Efficiency 'gains' used to reduce staffing levels, leading to increased stress and staff turnover dues to disaffection.
Transparency—doing what they say.	Regulated community knows too much about the priorities and tactics of the regulator.
There is clarity on the resource allocation across the different sectors so that industry fees and charges are allocated more fairly.	Resources taken up in national coordination and reporting functions, and less time is spent directly regulating industry.
Industry and regulator have more opportunity to engage on agreed priorities resulting in progress on issues previously left unresolved.	There is a risk of 'regulatory capture' as senior regulators are softened to the industry line which in turn may undermine the actions of local inspectors.
Improved capability and development opportunities for staff to enhance their skills and influence national decision-making.	Tension between local managers and national sector team where the most capable staff are in demand. This is especially difficult in a period of constrained public services.
Knowledge transfer through central training and 'communities of practice' networks to further best practice, innovation and promotion of consistency in regulation.	This may encourage hobbyists and lead to less time on direct regulatory activities.
Provides the ability to test new 'alternative' and complementary approaches to regulation that require national coordination and stakeholder engagement. The trial of environmental management systems (EMS)-based compliance and self-assessment is one example.	Local officer time may be diverted away from day job regulation to providing support to national trials that may or may not lead to any regulatory benefit.
Provides added value and opportunities for joined-up regulation as related regulatory initiatives (e.g., in water, energy, pollution prevention, climate change, etc.) can be promoted and brought to the attention of regulated industries.	Time taken up on non-'core' and non-income-generating activities may dilute the time spent on regulation.

The findings and lessons learnt from the project are expected to inform the way that regulators Account Manage larger regulator customers and the application of data standards. The aim is to improve the quality of the data regulators collect and their capability to use it for strategic assessment of business performance.

(iii) *Open data*

Government departments, regulators included, are increasingly making their data available online. Most of these data-sets are available free or obtainable with a licence (free or for a modest fee for large complex data-sets). The principle is to make as much data available as possible to help promote innovation and to encourage the use and re-use of governmental data. Key data-sets available from the UK regulators include details of permitted sites and activities, their environmental releases (pollution inventory), waste production data, information on regulatory compliance and prosecutions. Environmental data-sets can be found on the UK government's data-portal.

(iv) *Guidance simplification*

Other authors have ably covered this subject, but suffice to say the streamlining of guidance has been a recurring theme for successive governments, and the results have had a significant impact on the delivery of regulation across the UK. This has resulted in the reduction in the level of guidance, including at the sector level. This increases the potential for risk of misinterpretation and inconsistency of regulation.

5.8 Summary, Conclusions and Latest Developments

The sector-based approach has evolved over the last 10 years and now provides a sound framework for regulators to organise, present and deliver efficient and cost-effective regulation. The benefits of the sector approach, however, are largely anecdotal, as there is little if any independent evaluation of the approach. It is incumbent on the regulators to demonstrate the benefits to stakeholders, including industries who fund the activities through fees and charges.

One of the key themes for UK environmental regulators in recent years has been the pursuit of simplification in their approach to regulation. This has been driven not only by the pressure of reducing budgets but also by the increasing demand from customers, both in industry and the political masters in government, who seek clarity and consistency on regulation and who rail against so-called gold-plating.

Industry Trade Associations tend to work in their sector channels or 'silos' and constantly expect to receive information and services that are tailored to their needs. EU and UK regulation has by and large reflected the sector classification of Industry, as illustrated in the Listed Activities of the Industrial Emissions Directive, for example. This provides a logical starting point for the organisation of regulatory activities.

One of the key tests of better regulation is transparency. However, the effect of austerity, with the EA, for example, having endured substantial staff reductions in recent years, has led to a reduction in strategic head office roles, such as proactive communication, engagement and evidence roles. This may marginally reduce costs to the charge payer, but, as industry face frustrations and delays in the service they receive, it does little to foster trust and the development of close working relationships between the regulators and their regulated communities.

Recently, the EA has taken to publishing high-level 'Evidence summaries' and pictorial sector strategies for the key sectors with supporting performance data. But often the information provided is so high level that it is hard to gain anything other than a strategic overview from the information. More needs to be done to make the raw data more widely available, through extending the number of data-sets accessible through the 'open data' programme, for example. Publication of industry performance data and compliance ratings would be an effective way of encouraging improvements in compliance.

In terms of wider-sector initiatives and partnerships, a wealth of expertise, case studies and guidance has been built up through the work of partner organisations and programmes. In many cases, this has supported the delivery of environmental outcomes required by regulation such as resource efficiency and pollution prevention objectives. And while some initiatives are ongoing, in recent years, there has been a significant reduction in the number of projects and a streamlining in the level of service provided. This is most probably a reflection of the reductions in grant funding and government cuts. In the future, it will be even more difficult to get such initiatives off the ground as EU funding will no longer be available.

Moving forward, the regulators across the UK have endured a period of substantial change. Change is an ongoing process, but now is the time for stability and for the approach to be fine-tuned and lessons learned from implementation. Operational staff and industry stakeholders, as well head office staff and politicians need to be involved in this process so that the desired outcomes can be achieved and the environment made a better place for everyone.

Appendix 1

Appendix 1a: Environment Agency: 'Food & Drink Sector Key Statistics 2015'

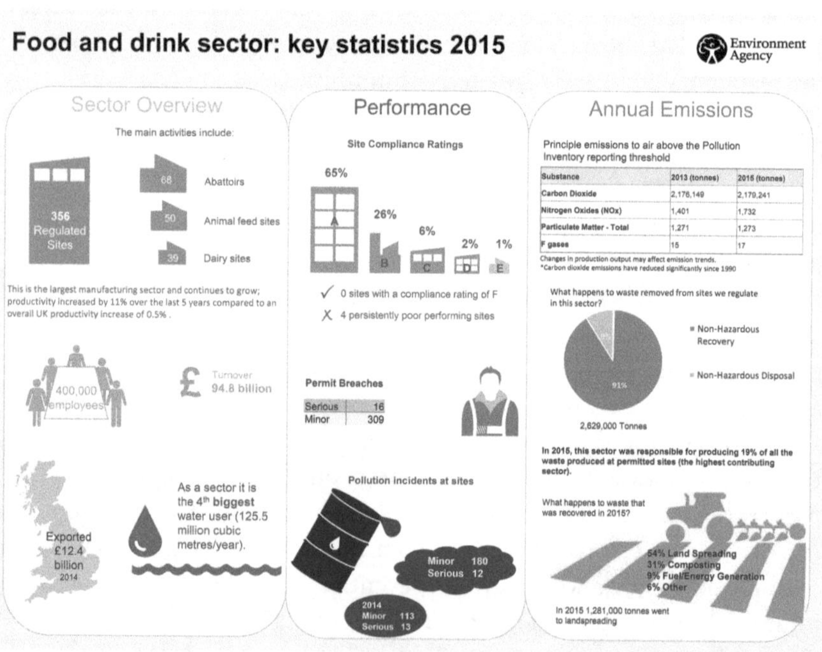

Appendix 1b: Environment Agency: 'Sector Strategy 2016–2020'

Food and drink: sector strategy 2016-2020

Environment Agency

Corporate Outcome

We will support sustainable growth by working with businesses to help them comply with regulatory requirements, whilst continuing to improve the environment and public health. Measuring the impact of their activities using compliance assessment, data and intelligence, holding them to account where necessary but helping those who are trying to do the right thing.

Objective	Intervention
Implement Industrial Emissions Directive (IED) in line with the adoption of Best Available Techniques Reference (BREF) documents.	Undertake sector review upon adoption of revised BREF(s). Measured by number of permits reviewed and new Best Available Techniques requirements implemented.
Work with industry to minimise amenity impact including all that are relevant (primarily odour).	Focus on priority sub-sectors. Measured by number of designated Sites of High Public Interest sites and length of time of designation (link to our existing Key Performance Indicator).
Encourage operators to adopt the waste hierarchy and achieve zero waste to landfill.	Promote resource efficiency for energy, water and waste. Provide advice and guidance to consumers/waste producers. Measured by waste trends from Pollution Inventory data (disposal to landfill and percentage of waste recovered).
Maximising compliance and taking tough action where permit conditions are not met. And ensure consistently high performance across the sector.	Use a risk based intelligence led approach for compliance and enforcement activities. Primary measure: total number of DEF performers including persistent poor performers.
Ensure industry minimises the environmental impact of refrigerant gases.	Check compliance with relevant legislation and take prompt action where necessary.
Greater resilience to climate change.	Provide advice and guidance on mitigating the effects of climate change.
Support sustainable growth by working with businesses to help them comply with regulatory requirements, whilst continuing to protect environmental and public health.	Applications for new permits, variations and transfers are processed on time. The sector contributes to achievement of National Emission Ceiling Directive targets for Air Quality.

Data Sources:
The Environment Agency holds records for the number and type of regulated sites, the number of pollution incidents and permit breaches, and the compliance rating for each regulated site. The number of pollution incidents include both permitted and unpermitted sites in this sector.
Substance release and waste transfer data is collected from sites that operate under a Part A(1) environmental permit and form part of the Environment Agency's Pollution Inventory. Details of the type and amount of waste received is collected from regulated sites which undertake waste activities under the Waste Framework Directive.
Industry data has been collected from a variety of sources including industry trade organisations and publicly available resources.

Appendix 2: Environment Agency Model 5-Year Strategy (Based on 2011–2015 Template)

Long-Term Outcomes for the xxxx Sector

This document sets out the key issues to be addressed during the period 2011–2015 through our permitting and compliance activities.

Sector Structure and Economic Overview

Key players and activities:	There are X operational permitted sites comprising (provide detail) Detail of sector breakdown
Total turnover:	Economic summary
Barriers to entry/ competition:	Impact of national/global markets and competition, technology barriers
Relevant trade associations:	Trade association(s) and key players
Performance banding:	A/B
OPA ratings, date	C/D
	E/F
Performance profile:	A/B
OPRA compliance ratings	C/D
	E/F

Regulatory Overview

The role of the Environment Agency:	Regulatory regimes that are applicable, for example, EPR installations, waste, EU-ETS, water abstraction, producer responsibility
Future regulation/changes:	Sector permit review due: Changes to regulation known or expected:

Community Issues

Local:	Outline local odour and amenity issues Sensitivity to climate change, for example, water resources availability and adaptation
National:	Any major issues with regulated customers. Proposed account management
International:	E.g., international regulation/agreements

Corporate strategy outcomes	Sector outcomes	Targets	Completion year				
			1	2	3	4	5
<List all Strategic outcomes which are relevant to the sector>	<List all outcomes—ensuring linkages to the Corporate Strategy and sub-strategies>	<Insert SMART target> (where available) or insert Corporate Strategy SMART target					
Compliance with environmental regulations (including permit conditions) has improved	Sector compliance objectives are assured through Industry participation in annual compliance reporting			✓			
Fewer sites have band D, E or F Opra performance scores	All operators have improved compliance with permit conditions (CS 2.4a)		✓	✓	✓	✓	✓
We receive fewer complaints relating to nuisance such as odour, dust and nuisance	All operators engage actively with their communities and complaints are reduced (Cs 2.4e)		✓				
More FTSE All-Share companies disclose their environmental performance	Major players proactively report their environmental performance and demonstrate improvements in key indicators						✓
Businesses we regulate use resources more efficiently and send less waste to landfill	We undertake a review of energy from waste and land spreading for the sector which informs investment decisions			✓			
	Resource efficiency (waste) production indicators improve year on year (CS 4.3a)		✓	✓	✓	✓	✓

(continued)

(continued)

Corporate strategy outcomes	Sector outcomes	Targets	Completion year
Businesses adapt to climate change	All sites have a plan that addresses increased flood risk, water abstraction pressures in drought-stressed areas and Effluent Treatment Plant operation safeguards for extreme weather events.		✓
Water use in industry is reduced	Following our audits, water use and water discharges are reduced and more sites are within BREF benchmarks for water use		✓
Over-abstraction within water bodies is reduced and fewer abstractions cause environmental damage	The environment is protected as abstraction from sensitive locations is reduced (CS 2.1d) (subject to further analysis of abstractions in sector)		✓
Greenhouse gas emissions from sites we regulate are reducing in line with government targets	Energy recovery, low carbon technologies and CHP are incorporated into company business plans where appropriate		✓
More regulated sites have acceptable environmental management systems	Regulated sites have improved environmental management systems, and we have more confidence in their regulatory performance		✓
All permits are delivered in time and to the required standard	The sector permit review is delivered to plan and achieves environmental and regulatory improvement (CS 2.5c)		✓

(continued)

(continued)

Corporate strategy outcomes	Sector outcomes	Targets	Completion year	
We improve how we work and administrative costs to businesses are reduced	Our regulation is more effective and nationally risk based with audits and inspections delivered through a coordinated audit programme.			✓
	We reduce industry reporting costs by simplifying data reporting requirements (CS 2.5b) (achieved through use of EPR sector permit template in permit review and the data simplification project)		✓	✓
	We act with one voice in dealing with customers and are able to take account of non-site-based regulatory issues, for example, emissions trading, packaging and broker appropriate solutions			✓
	We revise the sector guidance following publication of the revised BREF review. Industry participates in the review and provides feedback through the new guidance hub			✓

References

Defra. 2011. WR1403 *Business Waste Prevention Evidence Review. Report L2m4-7 Waste Minimisation Clubs.* November. Oakdene Hollins, Brook Lyndhurst, Resource Recovery Forum.

Department for Business Innovation and Skills, Better Regulation Delivery Office. 2014. *Draft Guidance: Non-economic Regulators: Duty to Have Regard to Growth.* January. ENTRUST http://www.entrust.org.uk/landfill-community-fund/statistics

Environment Agency. *Annual Reports and Accounts.* https://www.gov.uk/government/collections/environment-agency-annual-reports-and-accounts

Environment Agency. *Board Meetings, Minutes of Open Meetings.* See Environment Agency governance information at https://www.gov.uk/government/organisations/environment-agency/about/our-governance

Environment Agency. 2006. *The Sector Plan for the Waste Management Industry.*

———. 2011. *Future Approach to Regulation. Programme Board Presentation.*

———. 2012. *Nuclear Sector Plan—Environmental Sector Plan for the Nuclear Industry*, Issue 3.

———. 2013a. *Regulating for People Environment & Growth.*

———. 2013b. *Sustainable Business Report 2012 Version 1.* November.

———. 2015a. *CLEAR Info Company Level Environmental Accountability Information and Reporting Laymans Report* (LIFE10 ENVUK175).

———. 2015b. *Regulating For People, Environment And Growth: 2014 Evidence Summary.* https://www.gov.uk/government/publications/regulating-for-people-environment-and-growth-2014-evidence-summary

———. 2016. *Regulating for People, the Environment and Growth 2015 Evidence Summary.* September. https://www.gov.uk/government/publications/regulating-for-people-environment-and-growth-2015-evidence-summary#history

Environment Agency Data Share. 2017. *Regulated Industry Sector Strategies 2016–2020.* March. https://ea.sharefile.com/share?#/view/sf4a50053b28440bb

Environmental Services Association and the Environment Agency. 2010 *Waste Sector Plan: 3 Year Progress Report 2006–9.* December.

European Commission. 2006. *Netregs Case Study 4.* ec.europa.eu/environment/archives/sme/pdf/netregs_en.pdf

House of Commons Hansard Written Answers. 2004 *HC Deb*, 13 February 2004, vol. 418, cc63-5W.

Irish EPA. *Reducing Hazardous Waste—The Hazred Project.* http://www.epa.ie/waste/nwpp/hazred/#.WK1tnOSmny0

Minerals Products Association 2011 Performance 2010. A Sector Plan Report from the UK Cement Industry. November.

Phillips, Paul S., et al. 1999. UK Waste Minimisation Clubs: A Contribution to Sustainable Waste Management. *Resources Conservation and Recycling* 27 (3): 217–247.

UK Cabinet Office. 2003. Better Regulation Task Force. *Principles of Regulation.*

UK National Industrial Symbiosis Programme. See www.nispnetwork.com/

United States Environmental Protection Agency. 1995. *Reinventing Environmental Regulation, National Performance Review.* President Bill Clinton and Vice President Al Gore. March.

WRAP. 2017. The Courtauld Commitment. March. http://www.epa.ie/waste/nwpp/hazred/#.WK1tnOSmny0

Jon Foreman is an independent environmental consultant and provides bespoke support and advice on the management of environmental and regulatory issues. Jon has over 30 years' experience working in senior national policy and operational roles with UK environmental regulators, engaging with governments and industry on environmental regulatory policy and practice in the UK, Europe and overseas.

In the early part of his career, Jon worked as a waste regulator and helped to pioneer environmental monitoring and control techniques for landfill and hazardous waste management. With the Environment Agency, Jon led the waste minimisation and prevention initiatives for over ten years before focussing on engagement and liaison with industry where he worked closely with the Food and drink industry, the Paper and Textile industries.

6

Implementing the Industrial Emissions Directive: The UK Environmental Permitting Regime for High-Risk Activities

Adrian Kesterson

6.1 Introduction

Industrial activities play an important role in the economic well-being of Europe, contributing to sustainable growth and providing high-quality jobs. However, industrial activities can also have a significant impact on the environment. The largest industrial installations emit a large proportion of the key atmospheric pollutants and have other important environmental impacts, including emissions to water and soil, generation of waste and the use of energy (European Commission 2007: 2).

Emissions from industrial installations have been covered by European Union (EU) legislation for many years. Since the 1970s, various directives have been drawn up which have ultimately led to the adoption of Directive 96/61/EC concerning Integrated Pollution Prevention and Control (IPPC). The IPPC Directive set out the main principles for the permitting and control of installations based on an integrated approach

A. Kesterson (✉)
CEnv, MIEMA, Expirea Environmental Consultancy Ltd, Nottingham, UK

J. Foreman (ed.), *Developments in Environmental Regulation*, Palgrave Studies in Environmental Policy and Regulation, DOI 10.1007/978-3-319-61937-8_6

and the application of Best Available Techniques (BAT) to achieve a high level of environmental protection, taking into account the costs and benefits. The so-called sectoral directives laid down specific provisions, including minimum emission limit values (ELVs) for the following industrial activities—large combustion plants, waste incineration, activities using organic solvent and titanium dioxide production (European Commission 2007: 2).

IPPC and the body of legislation on industrial emissions also had a part to play in addressing the calls from the Member States, the European Parliament and other stakeholders for "Better Regulation". With this in mind, the Commission launched a two-year review in 2005 of the body of legislation on industrial emissions in order to ensure its environmental and cost-effectiveness as well as to encourage technological innovation. Based on a detailed analysis of the current situation and the outcome of the review process, the Commission proposed to streamline and enhance its policy on industrial emissions (European Commission 2007: 3).

The streamlining of permitting, reporting and monitoring requirements as well as a renewed cooperation with Member States to simplify implementation were earmarked to reduce unnecessary administrative burdens of between €105 and €255 million per year across the EU (European Commission 2007: 3). The Commission put forward a package of measures designed to address specific problem areas, which, over time, it was hoped would lead to an improvement of the situation. One of the key initiatives envisaged in this context for streamlining was the revision of the current legislation on industrial emissions through simplification, clarification and strengthening (European Commission 2007: 4).

The impact assessment undertaken as part of this initiative indicated that the problems identified could not be addressed without some changes in the legislation which included the following (European Commission 2007: 5):

1. Re-casting the existing pieces of legislation (seven) into a single Directive on industrial emissions. This would improve clarity and coherence both for Member States and for operators, has the potential to reduce unnecessary administrative burdens through combined permitting and streamlined reporting requirements, and will also achieve some environmental benefits.

2. Improving and clarifying the concept of BAT to create a more coherent application of the current IPPC directive and by requiring decisions that set permit conditions outside BAT to be justified and documented. In addition, current minimum ELVs in some sectors (e.g. large combustion plants) are tightened to ensure the progress needed for achieving the objectives of the Thematic Strategy on Air Pollution.
3. Introducing minimum provisions regarding inspection, review of permit conditions and reporting of compliance. Incentives for eco-innovation and support for the creation of lead markets are also considered.
4. Extending the scope of the IPPC Directive to cover certain activities (e.g. combustion plants between 20 and 50 MW) and clarifying the scope for certain sectors (e.g. waste treatment) to increase consistency and coherence of current permitting practices.
5. Where the Commission is to take measures to amend non-essential elements of the re-cast Directive, the Commission will be supported by a Comitology Committee and will ensure broad involvement of stakeholders.

The proposal for a Directive on industrial emissions was adopted to improve the EU policy on industrial emissions in December 2007. The Industrial Emissions Directive (IED) (Directive 2010/75/EU) (IED) came into force on 6 January 2011 and served to re-cast seven existing Directives related to industrial emissions into a single clear and coherent legislative instrument. Re-casting brought together, into a single new act, a legislative act and related acts, and passed through the full legislative process and repealed all the acts being re-cast. The re-cast included the IPPC Directive, the Large Combustion Plants Directive, the Waste Incineration Directive, the Solvents Emissions Directive and three Directives on Titanium Dioxide (European Commission n.d.).

The IED aims to achieve significant environmental and public health benefits by reducing emissions across the EU Member States, in particular, through better application of BAT. It introduced minimum provisions for the inspection of industrial installations, the review of permits, reporting on compliance and protection of soil. The scope of the legislation was also clarified and amended to include some new activities, in the waste treatment and food and drink sectors, for example.

The IED provides a single coherent regime that should help remove ambiguities and inconsistencies across Member States, promote cost-effectiveness and encourage technological innovation.

The IED was transposed into national legislation by Member States by 7 January 2013. In the UK, the IED was implemented through amendments to the following legislation:

- The Environmental Permitting (England and Wales) Regulations (EPR)
- The Pollution Prevention and Control (Scotland) Regulations (PPC)
- The Pollution Prevention and Control (Industrial emissions) Regulations (Northern Ireland) (PPC)

6.2 Prescribed 'Higher Risk' Activities

Most industrial and waste management activities have the potential to cause environmental damage to some extent, and there are general legislative controls for environmental aspects such as waste and water pollution. Some activities are inherently higher risk due to their nature, the hazardous properties of the materials used or the scale on which they are undertaken. 'Higher risk' activities whose operations could potentially have a significant environmental impact are prescribed in Annex 1 to the IED and are required to hold a permit in accordance with the IED. Annex 1 includes chapters of general activity descriptions which include sections and paragraphs covering more specific activities and qualifying criteria. This is similar to Annex 1 of the IPPC Directive, but a number of new activities were included, such as installations for the production of wood-based panels, new waste management activities, gasification or liquefaction, and new food and drink activities.

The chapters and examples of IED Annex 1 Prescribed Activities are shown in Table 6.1.

Schedule 1 of the EPR lists all the IED Annex 1 activities, and these are called "Part A activities". Most of these activities were previously listed in the IPPC Directive (Directive 2008/1/EC) (which originally came into force in 1996 and was repealed in January 2014) (Environment Agency 2015: 1).

Table 6.1 Chapter descriptions of the industrial emissions directive

Chapter	Description	Examples
1	Energy industries	Combustion of fuel
2	Production and processing of metals	Melting and refining of metals
3	Minerals industries	Cement and lime manufacture
4	Chemical industry	Inorganic chemicals manufacture
5	Waste management	Incineration of waste and waste treatment
6	Other activities	Paper manufacture, food and drink industries

Only in England and Wales the Part A activities are split into A(1) and A(2) activities. The A(1) activities are regulated by Natural Resources Wales (NRW) or the Environment Agency (EA) in England, and the lower-risk A(2) activities are regulated by Local Authorities. Elsewhere in the UK, the prescribed IED Annex 1 activities are all Part A activities under the PPC Regulations in Scotland and Northern Ireland, and are regulated by the Scottish Environmental Protection Agency (SEPA) and Department of Environment Northern Ireland (DOENI), respectively.

Prescribed activity descriptions often refer to "capacity" (or equivalent terms), for example, paragraph (a) of Chap. 3, Mineral Industries, Sect. 3.1, Production of Cement and Lime:

> Producing cement clinker in rotary kilns with a production capacity exceeding 500 tonnes per day or in other kilns with a production capacity exceeding 50 tonnes per day

The European Commission provided guidance on interpretation and determination of capacity under the IPPC Directive, and this is still applicable to the IED. The UK Regulators adopted this guidance and the EA's Regulatory Guidance Series, "RGN 2, Appendix 1—Interpretation of Schedule 1 to the Regulations, Appendix 2—Defining the scope of the installation" is a useful reference when assessing if an activity is capable of being operated above a prescribed threshold.

In all cases, capacity refers to potential capacity and not historical or actual production levels or throughput. This means that the design capacity of the installation will usually be the key issue, and this should be based on the maximum operating hours possible. Technical restrictions can influence capacity, such as chemical reaction or firing times, or essential cleaning stages between batches. Regulatory or other limits, such as restrictions on operating hours can also be considered (Environment Agency 2015: 2).

Any installation undertaking a prescribed activity at or above the capacity, or other qualifying criteria specified for that prescribed activity will require an environmental permit in England and Wales or a PPC permit in Scotland and Northern Ireland.

6.3 Permit Application Requirements

Article 11 of the IED requires Member States to take the necessary measures to ensure that permitted installations use all the appropriate preventive measures against pollution, apply the BAT and cause no significant pollution. The requirements of Article 11 are addressed in the permit application criteria prescribed in Article 12 of the IED, and these are transposed into the UK legislation, for example, in paragraphs three and four of Schedule 7A of the Environmental Permitting Regulations (EPRs; England and Wales).

Applications for permits must include details of the quantities, monitoring and potential impact of these emissions. Measures to be taken for the restoration of the site to remove the risk of pollution upon cessation of the activities must also be addressed. Details of the installation and its activities, including a non-technical summary are also required and, in particular, the proposed technology and other techniques for preventing or, where this is not possible, reducing emissions from the installation. The applicant must discuss the main alternatives to the proposed technology, techniques and measures studied.

The environmental scope of the Article 12 permit application criteria is summarised in Fig. 6.1.

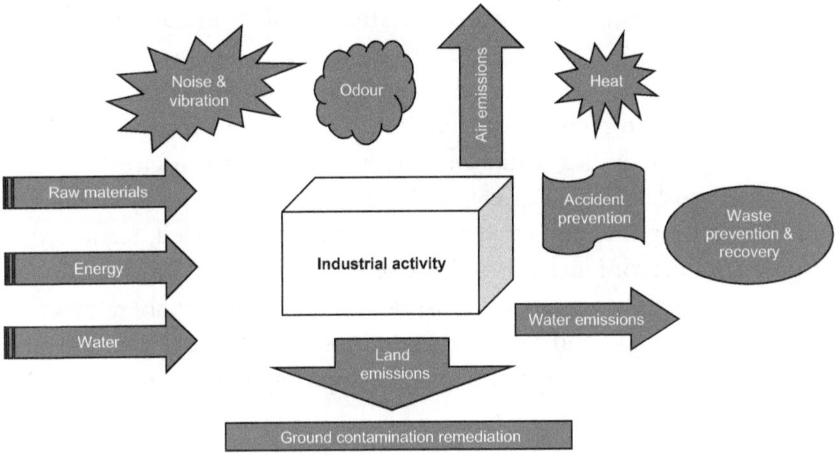

Fig. 6.1 Article 12 permit application criteria

6.4 IED Links to UK Legislation

European Directives (directives) and international agreements contain a variety of requirements, some of which can be delivered through a permitting and compliance system, and some of which are delivered in other ways. The majority of environmental quality and specific permitting standards, and other related requirements for environmental and human health protection come from directives. The EPRs, for example, ensure that those directives, national policy requirements and outcomes that can be delivered through a permitting and compliance system are delivered by the regime. The EPR places duties on regulators to exercise their permit-related functions to deliver the obligations and outcomes required by the relevant directives and, in some cases, national policy. In practice, this means that the regulator will ensure, where a permit is granted, that permit conditions achieve the objectives and intended outcomes of any of the directives or national policy which apply. The Regulations also give regulators powers in relation to their permit-related functions. The EPR Schedules generally identify particular requirements, (usually Article by Article, in the case of directives), which must be delivered through the permitting system. In some cases, requirements to be delivered through the permitting system are located in other legislation (Department for Environment, Food and Rural Affairs 2013: 5).

The IED links to other UK/EU legislation and allows permit conditions to be set to implement the requirements of the associated legislation for prescribed activities, while not prejudicing the requirements of the associated legislation. In Article 22, for example, the generation of waste must be prevented or else waste should be re-used, recycled, recovered or disposed of in accordance with Directive 2008/98/EC (Waste Framework Directive). This Directive was transposed into UK national laws such as The Waste (England and Wales) Regulations 2011. In this case, a typical permit will require the operator to take appropriate measures to ensure that:

(a) The waste hierarchy referred to in Article 4 of the Waste Framework Directive is applied to the generation of waste by the activities.
(b) Any waste generated by the activities is treated in accordance with the waste hierarchy referred to in Article 4 of the Waste Framework Directive.

Similar links can be found in Article 22 of the IED which specifies the site closure requirements to ensure that any contamination of the ground or groundwater from prescribed activities is remediated upon cessation of activities. Permit conditions are set by regulators to cover these requirements but should not prejudice the requirements of the Water Framework Directive (WFD) (2000/60/EC) and the Groundwater (WFD) (England) Direction 2014. The Environmental Liability Directive 2004/35/EC also relates to this and is transposed in UK law under the Environmental Liability (Scotland) Regulations, for example.

6.5 Determination of Best Available Techniques

The IED's integrated approach means that permits must take into account the environmental performance of the plant as a whole, covering the aspects required under Article 12. The permit conditions, including ELVs, must be based on the BATs, as defined in Article 3 of the IED. These definitions are mirrored in the UK implementing legislation and are summarised in Fig. 6.2:

To help authorities, businesses and other interested parties determine BAT, the European Commission organises an exchange of information between experts from the EU Member States, industry and environmental organisations. BATs are determined by a Technical Working Group steered by the European IPPC Bureau (EIPPCB) at the EU Joint Research Centre (JRC). This process results in a Commission Implementing Decision establishing BAT Conclusions for each sector which are laid out in the BAT Reference Document (Bref).

The BAT Conclusions play a central role in the implementation of the IED and lay down their description, information to assess their applicability, the emission levels associated with the BATs, associated monitoring, associated consumption levels and, where appropriate, relevant site-remediation measures. The BAT Conclusions are the reference for competent authorities when setting permit ELVs which must ensure that, under normal operating conditions, emissions do not exceed the emission levels associated with the BATs as laid down in the BAT Conclusions (Official Journal of the European Union 2015: 1).

There are currently 33 Brefs covering the industry sectors prescribed in the IED as well as generic aspects such as energy efficiency, and they are

Best	Most effective in achieving a **high general level** of protection of the environment **as a whole**
Available	Developed on a scale which allows implementation in the relevant industrial sector, under **economically and technically viable conditions**
Techniques	**Both** the technology used and the way in which the installation is **designed, built, maintained, operated and decommissioned**

Note: in determining BAT, special consideration should be given to the criteria listed in Annex III of the IED

Fig. 6.2 Definition of BAT in the IED

reviewed around every eight years. From 2012 to 2020, the commission will publish BAT Conclusions for each industrial sector, which will specify BAT Associated Emissions Levels (BAT-AELs). Within four years of the publication of BAT Conclusions, regulators must review the permits in that sector and set ELVs that are based on the BAT-AELs. However, Article 15 of the IED allows Regulators to grant derogations under certain circumstances and set ELVs that are less strict than the BAT-AELs, for example, where it can be shown that meeting the levels set out in the BAT Conclusions is not feasible, as this would lead to disproportionate costs versus benefits.

In March 2012, the European Commission published the first BAT Conclusions for the Iron and Steel and Glass industries, so the permit reviews in these sectors should be completed by March 2016. The Commission has not published guidance on the determination of derogation requests. Each Member State is developing its own proposals for implementation which could lead to varying interpretations. In 2014, the first European Network for the Implementation and Enforcement of Environmental Law (IMPEL) derogations project provided an opportunity to compare the proposed approaches to IED permit reviews and derogations. It recommended running a second project in 2016 to compare how well these approaches have worked in practice. The report from the second project is expected late in 2016 (IMPEL 2016a: 1).

BAT Criteria and Environmental Quality Standards

The BAT approach complements, but differs fundamentally from regulatory approaches based on Environmental Quality Standards (EQS). Essentially, BAT requires measures to be taken to prevent emissions and measures that simply reduce emissions and are only acceptable where prevention is not practicable. Thus, if it is economically and technically viable to reduce emissions further, or prevent them altogether, then this should be done irrespective of whether or not EQSs are already being met.

The BAT approach requires us not to consider the environment as a recipient of pollutants and waste, which can be filled up to a given level, but to do all that is practicable to minimise emissions from industrial activities and their impact. The BAT approach first considers what emission prevention can reasonably be achieved and then checks to ensure it is met. The BAT approach is, therefore, the more precautionary one because the release level achieved may be better than that simply required to meet an EQS.

Conversely, if the application of the BAT might lead to a situation in which an EQS is still threatened, additional measures shall be included in the permit to comply with the EQS. This scenario essentially requires the installation to go beyond BAT. Article 18 of the IED allows for expenditure beyond BAT where necessary, and, ultimately, an installation will only be permitted to operate if it does not cause significant pollution.

6.6 Assessing BAT at the Installation Level

When assessing the applicability of sectoral BAT standards at the installation level, the selection of the most appropriate technique(s) may depend on local factors and, where the answer is not self-evident, an installation-specific assessment of the costs and benefits of the available options will be needed. Most UK Regulators still encourage permit applicants to use the EA's "H1 Environmental Risk Assessment Framework" and its associated software tool to help with the assessment.

H1 was withdrawn by the EA in February 2016, however, and has been replaced by "Guidance on risk assessments for your environmental permit". This is a more generic, qualitative type of environmental impact assessment to identify whether releases to air, water or land are a risk to the environment and whether a more detailed modelling assessment is required.

Risk assessments must demonstrate how an applicant's activities will be managed so that their impact on their local environment is acceptable to the Regulator. Various software modelling tools are available to quantify the environmental risk posed by activities against a series of defined parameters and calculate the impact of proposed substance releases to various media. These can be used to screen out from further detailed assessment any 'insignificant' emissions to air, depositions onto land or discharges to water where they are not "liable to cause pollution".

As mentioned earlier, the IED allows for derogations from the BAT Conclusions if certain conditions can be demonstrated. To benefit from such a derogation, operators must apply to the Regulator and demonstrate that BAT would lead to disproportionately higher costs compared to the environmental benefits due to the geographical location, or the local environmental conditions of the installation concerned or the technical characteristics of the installation concerned. Any derogation requires public consultation, and the rationale for denial must be robust in case of any appeal. This is likely to place a large burden on regulators, and the associated costs must be recovered via charges to permit holders. The EA Charging Scheme treats this work in respect of facilities regulated by them as the equivalent of a substantial variation for the purposes of charging. However, Local Authority fees for A2 permits in England do not currently cover such work and DEFRA are reviewing the charging scheme to cover this in 2017 (DEFRA 2016: 4).

The Regulators should encourage the development and introduction of innovative techniques that advance BAT standards criteria. This means techniques which have been developed on a scale which reasonably allows implementation in the relevant sector, which are technically and economically viable and which further reduce emissions, and their impact on the environment as a whole.

6.7 Implementation of the IED Across the EU

The IED is implemented in the national legislation of Member States and a permitting regime is required in each country. The benefit of the IED is that it should remove enforcement and implementation inconsistencies across the Member States. This can only improve the sustainability of industry. In the past, different approaches to environmental regulation across Europe have penalised UK industry, where there has not been a level playing field. As an EA Inspector, I recall a chemical manufacturer I regulated telling me that their PPC permit was much more onerous than one they held for a site they operated in Italy. The Italian permit was essentially just a one-page licence listing the company name, address and the prescribed activities undertaken.

The UK animal by-products processing industry has been disgruntled for many years about the requirement for the thermal oxidation of high-intensity odour imposed under UK permits being stricter than in some European countries where less energy-intensive techniques are used. The industry remains sceptical that BAT will be applied consistently across the EU under the IED. The review of the Bref for the animal by-product industries sector is expected to start in 2017 and is an opportunity to assess the different techniques and standards across the EU, and set the BAT Conclusions for implementation in permits across the EU.

IMPEL has worked on facilitating the implementation of the IED into the day-to-day work of competent authorities. Their projects, such as IED inspections and permitting, have paved the way for a better understanding and enforcement of the IED obligations across the EU. This is a further step forward towards the establishment of a level playing field concerning the implementation of the IED within Europe. IMPEL is preparing annual technical workshops for a bigger group of participants to present and discuss the approaches of national authorities and exchange good practice. Joint-site inspections will be organised and the implementation of the IED will be supported. In parallel, new tools to support permitting and inspection are being developed by IMPEL (IMPEL 2016b: 1).

6.8 Environmental Permitting in the UK

Permitting of "higher risk" activities is not a new regime in the UK. Some established industry sectors, such as combustion and chemicals, have been required to hold a permit for many years under PPC and before that authorisations under the IPC regime. Other sectors such as some waste management activities are relatively new to this integrated type of permitting regime.

The EPRs (England and Wales) 2010 were introduced before the IED to create a generic permitting regime that could be applied consistently to the permitting of prescribed activities, waste management operations, radioactive substances, water and groundwater discharges. All permit applications are determined by the National Permitting Services (NPS) in England and Wales. This central permitting structure functions well for the EPR as permit applications are more generic, less complex, and standard permit templates and conditions can be used. This has improved permit determination times and has led to more consistent permits and permitting decisions. The downside of this approach is that the local regulatory knowledge of an installation is not always addressed during permit determination as NPS staff are usually unfamiliar with the installations they are permitting.

Although the EPR has streamlined permitting, the reduction in governmental department resources in recent years has led to useful regulatory and industry sector guidance being discontinued. Combined with staff cuts and the growing shortage of experienced regulators, there is a danger that industry and regulators will not fully understand the legislation. Industry may struggle to apply for and comply with their permits, and the regulators could struggle to enforce them. This is a frustration for both industry and regulatory staff and is a familiar scenario I experienced during my time in industry regulation in Western Australia where these problems are more extreme.

Some UK industry sectors remain pessimistic about consistent application of the IED in the UK compared to Europe. Different interpretations, such as prescribed activity capacity are a concern, and this can result in some installations requiring permits or paying higher fees than similar sites elsewhere in EU.

Fees and charging schemes for permitted installations are another variable across the EU. Operators of prescribed activities are required to pay fees to the regulator to cover the regulator's costs for permitting and regulating the installation. This is in line with the "Polluter Pays" key principle, where the operator bears the cost of any measures to prevent environmental harm that they may cause.

The permit application and annual permit fees in England and Wales are based on a risk-based profile which address a range of factors such as scale, complexity, location, emissions, management systems and compliance performance. This Operational Risk Appraisal (OPRA) scheme was introduced around 2004 for the prescribed activities under PPC and has evolved to cover all permits under EPR. The concept of OPRA is that the higher risk, most polluting, poorly managed installations require more regulatory effort and so should pay higher fees. This offers incentives to improve compliance performance and reduce emissions. OPRA is very complex, however, and errors in fee calculations at the time of permitting are often carried over in subsequent years despite annual OPRA reviews.

The Environmental Regulation (Scotland) Charging Scheme 2016 has replaced five previous charging schemes with a single risk-based scheme. These changes aim to improve environmental regulation and link regulatory effort and charges to environmental risk. In Northern Ireland, the charging scheme is still based on a component system that was previously used elsewhere in the UK. This format applies fees to components allocated to different process stages of the prescribed activities and does not take account of the operator's performance and levels of emissions.

Regulators use OPRA profiles in their business planning to target their regulatory effort and focus on poor-performing installations. Data analysis can be used to identify the most common sectors, permit conditions or type of event causing non-compliance, such as spillages in the anaerobic digestion sector due to poor management systems and training. The Regulators use this information to plan national auditing campaigns to drive improvements where the risks are highest, such as recent audits of bunding at hazardous waste storage installations.

The compliance data used in OPRA come from the Compliance Assessment Reports (CARs) regulatory officers completed during site inspections, audits and reviews of reports. Any non-compliances with permit conditions are recorded on CAR forms and entered into a Compliance Classification System (CCS) database. Non-compliance ratings range from 1 to 4, with category 1 being the most severe, with actual or potential for major environmental impact. Category 2 and 3 events have lower environmental impact, and category 4 is for failures in submitting permit reports where there is no environmental impact.

Officers can refer to various non-compliance examples in their CCS guidance with suggested compliance ratings. Inconsistency across Officers is inevitable, however, and in some cases, Officers within the same local team struggle to classify a particular example the same. This is an important issue as less experienced and pragmatic Officers can accumulate numerous non-compliances in one or two visits, often for the same event.

This is contrary to the CCS guidance in some cases and leads to a disproportionally large non-compliance score and larger fees. Some operators set their staff targets for permit compliance, and staff who may lose financial bonuses for non-compliances are very sensitive on this issue and occasionally challenge non-compliance decisions.

Another feature of the UK approach is that companies operating across different countries in the UK may be regulated by two or more regulators, the EA, NRW, SEPA or DOENI. This can lead to inconsistencies in permits for the same prescribed activity and can give competitors an advantage, where they have less onerous permit conditions or regulatory interpretation. As common guidance such as H1 is withdrawn, these inconsistencies may become more profound. This is a challenge for the UK as the mandatory BAT Conclusions are rolled out under the IED.

6.9 Successes of the IED

Article 2 of the 7th EU Environmental Action Programme (EAP) "Living well within the limits of our planet" lists nine priority objectives for 2020, several of which are particularly relevant to Industrial Emissions policy, such as:

(a) to turn the EU into a resource-efficient, green and competitive low-carbon economy
(b) to safeguard the EU's citizens from environment-related pressures and risks to health and well-being
(c) to maximise the benefits of the EU's environment legislation by improving implementation

Under the current Commission, industrial emissions policy is framed by the priority areas such as jobs, growth and investment, a forward-looking climate change policy and the creation of a strengthened industrial base (European Commission 2014: 1).

During 2015–2016, the Commission is identifying and documenting "success stories" resulting from the application of the IED and its predecessor directives. These successes from the last decade are to be primarily judged in terms of environmental and health achievements and, where possible, in conjunction with economic and social achievements relating to priority areas. For example, to inform the methodology, the Commission looked at the potential environmental benefits of the decommissioning or conversion of mercury cell plants in the chloro-alkali sector as a consequence of industrial emissions policy (IPPC Directive and IED). The main benefits of such a conversion will be reduced emissions of mercury to air and water, reduced generation of waste contaminated with mercury and reduced electricity consumption.

Based on the 2010 reference data, the total annual mercury emissions to air of the chloro-alkali industry in the 27 countries of the EU (EU-27) amounted to about 6 tonnes or approximately 7% of the total anthropogenic mercury emissions to air. These emissions would be avoided by complete decommissioning or conversion of all mercury cell plants. In 2010, the total electricity consumption of the chloro-alkali sector in the EU-27 and European Free Trade Association (EFTA) countries amounted to 35 TWh. This was equivalent to 1% of the total final energy consumption in the form of electricity in this region. In 2010, about one-third of the total chloro-alkali production capacity in the EU-27 and EFTA countries was based on the mercury cell technique. The conversion of a mercury cell plant to the membrane cell technique typically results in a reduction of electricity consumption of approximately 30%. From these figures, it can be estimated that a complete conversion of all mercury cell plants to the membrane cell technique would result in a reduction of 10% of the total electricity consumption of the chloro-alkali sector, equivalent to approximately 3.5 TWh or 0.1% of the total final energy consumption in the form of electricity in the EU-27 and EFTA (European Commission 2014: 6).

6.10 What Next for UK Policymakers?

The future challenges for UK policymakers are likely to revolve around global and EU drivers such as the circular economy principles, sustainability, carbon footprinting, climate change and alternative energy supplies. The 2015 United Nations Climate Change Conference, COP 21 secured a commitment to keep global temperature rise "well below" 2 °C if parties sign up to this by April 2017. Similarly, the new global food waste standard, "The Food Loss and Waste Accounting and Reporting Standard" (or FLW Standard) encourages countries, companies and other entities to account for and report food loss waste. This could open up a major new market for green technologies.

It is encouraging that the UK is consulting industry on its plans to revamp its energy efficiency schemes and associated industry incentives to accommodate the UK's commitments towards its more expensive targets for renewable energy. This is important so as to not penalise the most energy-intensive industries such as the steel industry.

On the other hand, the UK's proposals to scrap the Feed-in Tariff (FiT) for anaerobic digestion (AD) sites larger than 500 kilowatts of electrical power (kWe), as well as further reductions on the tariff support available for small and medium plants would see support drop to around a third of its 2014 level by 2017. Industry groups fear that this steep cut to the levels of support provided to new AD plants could severely hold back development of AD infrastructure in large parts of the UK.

Adapting legislation to tie in with these global and EU drivers while not discouraging the emerging technologies is a challenge. Policymakers will need to be mindful not to deter these emerging sectors with regulatory burdens and permitting restrictions.

References

Department for Environment, Food and Rural Affairs (DEFRA). 2013. *Environmental Permitting Guidance Core Guidance for the Environmental Permitting (England and Wales) Regulations 2010*. March. https://www.gov.uk/government/uploads/system/uploads/attachment_data/file/211852/pb13897-ep-core-guidance-130220.pdf. Accessed 20 May 2016.

———. 2016. *Review of Local Authority Environmental Regulation Fees and Charges—Call for Evidence*. September. https://consult.defra.gov.uk/industrial-pollution-control/la-epr-fees-charges. Accessed 24 Oct 2016.

Environment Agency. 2015. *Understanding the Meaning of Regulated Facility*. Regulatory Guidance Series, No RGN 2, Appendix 1—Interpretation of

Schedule 1 to the Regulations, Appendix 2—Defining the Scope of the Installation. May. https://www.gov.uk/government/uploads/system/uploads/attachment_data/file/435475/LIT_6529.pdf. Accessed 8 Apr 2016.

European Commission. 2007. *Communication from the Commission to the Council, the European Parliament, the Economic and Social Committee and the Committee of the Regions Towards an Improved Policy on Industrial Emissions.* December. http://eur-lex.europa.eu/legal-content/EN/TXT/PDF/?uri=CELEX:52007DC0843&from=EN. Accessed 17 May 2016.

———. 2014. *Identification and Documentation of Industrial Emissions 'Success Stories'.* https://circabc.europa.eu/sd/a/9f45a538-d92a-4a2f-aa18-0dd5025881b7/IED%20Success%20stories-ToR.pdf. Accessed 13 Jun 2016.

———. n.d. *New Proposal for a Directive on Industrial Emissions.* http://ec.europa.eu/environment/archives/air/stationary/ippc/proposal.htm. Accessed 16 May 2016.

IMPEL. 2016a. *Project 2016/02: Experience of Derogations from IED BAT-AELs.* April 7. http://www.impel.eu/wp-content/uploads/2016/04/IED-Derogations-Project-Abstract-EN_FR-April-2016.pdf. Accessed 25 Oct 2016.

———. 2016b. *Supporting IED Implementation.* http://www.impel.eu/wp-content/uploads/2016/04/IED-Implementation_abstract-English-version-April-2016.pdf. Accessed 25 Oct 2016.

Official Journal of the European Union. 2015. *Commission Implementing Decision (EU) 2015/2119 of 20 November 2015, Establishing Best Available Techniques (BAT) Conclusions, Under Directive 2010/75/EU of the European Parliament and of the Council, for the Production of Wood-Based Panels.* November 24. http://eur-lex.europa.eu/legal-content/EN/TXT/?uri=uriserv:OJ.L_.2015.306.01.0031.01.ENG. Accessed 24 Oct 2016.

Adrian Kesterson is Director of Expirea Environmental Consultancy Ltd. based in Nottingham, UK. He is Chartered Environmentalist, Full Member and Registered Auditor of the Institute of Environmental Management and Assessment (IEMA).

He is Technical Advisor for the Food-chain and Biomass Renewables Association (FABRA UK) and advises on animal by-products, environmental and energy issues at the UK and EU level.

He has prepared environmental permit applications and undertaken environmental legal compliance audits. He was Process Industries Inspector at the Environment Agency (EA) for 15 years, covering most industry sectors and was a member of the EA's Food and Drink Group.

7

Environmental Regulation for High-Risk Materials and Hazardous Wastes

Ken Westlake

7.1 Introduction

This chapter looks at developments in the regulation of higher risk activities, especially relating to more hazardous materials, including hazardous wastes. The development of UK and European controls has been divided across three key drivers:

- Incident response
- Global and more locally identified environmental concerns
- Environmental policy and strategy (including elements of each of the above)

Controls related to each of these will be examined, together with links to the circular economy and sustainability.

The UK has, for a long time, legislated against potentially polluting activities, with many of the very early controls relating to emissions from

K. Westlake (✉)
Westlake Associates, West Bridgford, Nottinghamshire, UK

J. Foreman (ed.), *Developments in Environmental Regulation*, Palgrave Studies in Environmental Policy and Regulation, DOI 10.1007/978-3-319-61937-8_7

high-risk activities and high-risk chemicals. The first example of such legislative control over more polluting industries was the Alkali Act of 1863, introduced to control acid gas emissions into the atmosphere from industrial manufacturing plants that were causing extensive damage to the environment and to human health. Controls under the Alkali Act sought to reduce emissions into air by introducing statutory emissions limits, and through the establishment of a central government body, the Alkali Inspectorate, to oversee the implementation.

The fact that the battle of Gettysburg was fought between Union and Confederate forces of the American civil war in the same year helps to put the level of control into some perspective.

A second Alkali Act of 1874 required those falling under its control to apply 'best practicable means' to manage polluting activities. The concept of Best Practicable Means (BPM) was only replaced for high-risk activities with the introduction of the similar concept of Best Practicable Environmental Option (BPEO) within Part I of the Environment Protection Act in 1990! This related to the then newly introduced concept of Integrated Pollution Control (IPC), of which more later.

The 1874 Act was followed by The Alkali Act of 1906 that also formed the basis for twentieth-century pollution control in Britain (Vogel 1986); it remained in place until the implementation of The Control of Pollution Act of 1974.

While the above demonstrates a long history of control over higher risk activities within the UK, the focus of what is to come will relate to more recent times, and will initially look at controls introduced in response to specific environmental incidents.

7.2 Incident Response

Within the UK, the early twentieth century saw little in the way of updated environmental legislation beyond that described above, and it was in response to a 'potential pollution incident' that further controls on high-risk activities were introduced; this occurred in 1972 when drums of cyanide waste were dumped at an abandoned brick kiln near Nuneaton, leading to a significant public outcry (Hansard 1972). At the time, there was no legislation that would allow anyone dumping such wastes to be prosecuted, and so The Deposit of Poisonous Waste Act 1972 was drafted in ten days as a private members' Bill and then quickly passed through the Parliament.

The Deposit of Poisonous Waste Act was introduced as a temporary measure, and in 1974 the Control of Pollution Act introduced further controls; amongst other things, these required those managing wastes to have a licence to operate, through which controls on day-to-day wastes management could be applied. This concept of waste licencing/permitting was adopted across Europe for the first time with the introduction of the Waste Framework Directive of 1975, and it provides a good example of where national (in this case UK) legislation has influenced European controls to provide for more effective management of environmental risks across all Member States.

The Control of Pollution Act represented a big step forward in the management of wastes, and although there were weaknesses in its content, it facilitated much more effective control over wastes management generally. At that time, wastes disposal did not differentiate between hazardous and non-hazardous wastes in the same way that we do today, and so controls on these more high-risk materials were, by default, also managed for better environmental protection.

More specific controls on higher risk wastes were introduced initially in 1978, when Council Directive 78/319/EEC (EEC 1978a) on toxic and dangerous waste was introduced for the management of higher risk wastes and related activities, and this is considered further under 'policy approaches' to the control of high-risk activities.

Given the level of control exercised in nineteenth-century Britain (e.g. The Alkali Act of 1863), it seems strange that requirements for waste licences/permits required the incident described above to precipitate more modern-day controls; this is most likely a reflection of a number of factors including 'World Wars', as well as changing industrial processes and technology development.

Beyond the UK, a number of incidents associated with high-risk activities have initiated changes to their management and control, with one of the more high-profile incidents being the one at Seveso in Italy in 1976, and because of which, there have been significant developments in the control of high-risk sites spanning the last three decades, and which continue to change and develop today.

The so-called Seveso Disaster was an industrial accident that occurred on 10 July 1976, at a plant manufacturing trichlorophenol in a small town, approximately 20 km north of Milan. Poor management control resulted in the accidental release of dioxin-containing emissions into the atmosphere that passed over the adjacent neighbourhood. In the immediate aftermath of the incident, large numbers of animals died, while some local residents suffered from skin lesions (Chloracne). It resulted in high exposure to the dioxin 2,3,7,8-tetrachlorodibenzo-p-dioxin (TCDD) in residential populations, in some cases, as high as 50 $\mu g/m^3$ (French Ministry of Environment 2008), but while many studies have been undertaken to evaluate long-term effects, the results remain inconclusive.

At the time of the incident, there were relatively few legal controls on the management of such high-risk activities, and no procedures were in place to deal with emergency situations; the local population was essentially unaware of the risks, while poor communication between the company and the authorities resulted in delayed actions that could have helped to minimise the adverse effects on human health and the environment (French Ministry of Environment 2008).

The Seveso incident changed this, and resulted in the publication of the original so-called Seveso Directive in 1982 (EEC 1982). This was updated as 'Seveso II' in 1996 (EC 1996a), and again as 'Seveso III' in 2012 (EC 2012a), and is implemented within the UK through the COMAH (Control of Major Accident Hazards) Regulations.

The original Seveso Directive comprised a three-part strategy that is still relevant today:

I. Identification of high-risk sites
II. The use of control measures to prevent major accidents
III. The use of mitigation measures to limit the effects of any accidents which do occur

Amongst other considerations, the first Seveso Directive required that all manufacturers should prove to the competent authority 'that they have identified existing major-accident hazards, adopted the appropriate safety measures, and provided the persons working on the site with information, training and equipment in order to ensure their safety'.

Relevant installations to which these new controls applied ('I' above), together with chemical storage limits and information to be supplied (where applicable), were identified within annexes to the Directive. This Directive represented a major step forward in site preparedness for, and the management of risks associated with, major incidents on high-risk manufacturing and other sites.

In 1996, and after a number of amendments of the 1982 Directive, a new Directive (known as 'Seveso II') on the control of major-accident hazards involving dangerous substances replaced the original Seveso Directive. Amongst other things, it sought to:

- Increase control in the light of changing European Policy to improve risk accident management
- Widen the scope of the Directive

This 'widening of scope' recognised that despite controls under the original Seveso Directive, major industrial incidents continued to occur, to the extent that Seveso II incorporated lessons learned from the catastrophic incident at Bhopal in India in 1984, in which thousands of people died as a result of the release of methyl isocyanate gas and other chemicals from a pesticide plant operated by Union Carbide (Box 7.1).

Box 7.1 The Bhopal Incident

In this incident, a cloud of methyl isocyanate gas enveloped the area surrounding a pesticide plant in Bhopal, India. By the time the incident was over, many were dead or injured. It has been reported (Varma and Varma 2005) that nearly 5000 people died within two days, and the death toll eventually reached 20,000.

In 2012, Seveso III amended Seveso II and was adopted especially recognising changes in the Union legislation (EC 2008) on the classification of chemicals (the so-called CLP Regulations), and increased rights for citizens to access information and justice, while also recognising the lessons to be learned from further industrial disasters, such as those at Toulouse and Enschede (Boxes 7.2 and 7.3).

Box 7.2 Explosion at Fireworks Factory, Enschede, Holland

On 13 May 2000, an explosion and ensuing fire occurred at a fireworks factory in Enschede, Holland. The company was located in the middle of a residential area, and the explosion and ensuing fire caused the death of 23 people (with approximately 1000 injured) and severely damaged or destroyed approximately 500 houses (van der Velden et al. 2009).

Box 7.3 Explosion at a fertiliser plant, Toulouse, France

On 21 September 2001, an explosion of ammonium nitrate at the AZF plant in Toulouse, France caused 30 fatalities (22 on site and 8 beyond the site boundary) and an estimated 2500 injuries. The explosion occurred in an ammonium nitrate store in which approximately 400 tonnes of the chemical were being stored (Barthelemy et al. 2001).

The disasters at Enschede and Toulouse were evidence that controls under the Seveso II directive were not sufficient to prevent major accidents turning into disasters. Moreover, a review of Seveso II showed that while it 'has been instrumental in reducing the likelihood and consequences of such accidents thereby leading to a better level of protection throughout the Union'. … 'A review of that Directive has confirmed that the rate of major accidents has remained stable' (EC 2012). Therefore, while significantly better protection was being afforded through controls under the Seveso Directive, industrial accidents continued to occur.

It is interesting to note that with both the incidents in Enschede and Toulouse, the plant was located in the suburbs of the city, and the accidents caused widespread damage both on and off-site. In this regard, Seveso III recognised that 'In order to provide greater protection for residential areas, areas of substantial public use and the environment, including areas of particular natural interest or sensitivity, it is necessary for land-use or other relevant policies applied in the Member States to ensure appropriate distances between such areas and establishments presenting such hazards and, where existing establishments are concerned, to implement, if necessary, additional technical measures so that the risk to persons or the environment is maintained at an acceptable level' (EC 2012a). In practice, many sites are located where they are for historical reasons (Box 7.4), and therefore protecting the general population in the event of a major incident can be difficult and requires special consideration. For this reason, land-use planning has been a key requirement of both Seveso II and III, and demonstrates the need for integrative approach to control whereby control of major-accident hazards is undertaken holistically, and where developments in time account for practical lessons that have been learned from experience.

Box 7.4 The relevance of site history to effective control in the twenty-first century

The history of the site in Toulouse (Box 7.3) dates back to the seventeenth century, when there was an explosives factory on the île de Tounis, which after a series of accidental explosions, relocated towards the South where it could continue to benefit from the energy provided by the river, whilst at the same time moving it further away from the growing city (Barthelemy et al. 2001). However, as both the plant and the city spread, due to World Wars and the need for explosives, and population growth, respectively, then they once again came to reside in close proximity. Modern-day controls introduced through the land-use planning element of the Seveso Directive would now limit such developments in a risk-assessed way, so that population risks were understood and managed through appropriate controls.

The Seveso III Directive now applies to more than 10,000 industrial establishments in the European Union, where dangerous substances are used or stored in large quantities in industries, such as the chemical, petrochemical, logistics and metal refining sectors.

Within the UK, the COMAH Regulations of 1999 (The National Archives n.d.-a) implemented the requirements of Seveso II but were revoked and replaced by the Control of Major Accident Hazards

Regulations 2015 (The National Archives n.d.-b), so that the requirements of Seveso III could be fully met. The land-use planning elements of the Directive are met through the planning legislation.

Changes and additions introduced under COMAH (2015) include:

- Substances covered by the Regulations have been updated and aligned to the 'CLP' (2008) Regulations (EC 2008)
- Transitional arrangements for safety reports have been introduced, together with
- A new requirement for co-operation by designated authorities in tests of the external emergency plan
- Stronger requirements for public information, including a duty for lower-tier establishments to provide public information
- Stronger requirements for competent authorities on inspection
- Local authorities must now inform people who are likely to be affected following a major accident

More generally, the COMAH Regulations place a duty upon those falling under their control to:

- Notify the competent authority, when specified dangerous substances on site exceed thresholds (Notification). The information to be provided is specified within the Regulations
- Obtain a hazardous substances consent (for those sites new to the controls) whereby this enables a hazardous substances authority (HSA) to consider whether the presence of a significant quantity of a hazardous substance is appropriate having regard to the risk to the community
- Prepare and work to a Major Accident Prevention Policy (MAPP)
- Operators must take all measures necessary to prevent major accidents, and limit their consequences to people and the environment

While for top-tier COMAH sites (those that present greater risk) additional requirements include:

- Preparation of a safety report to be sent to the Competent Authority to help demonstrate that all measures necessary have been taken to prevent major accidents, and to limit the consequences to people and the environment of any that do occur

- Prepare and test a site emergency plan to ensure that effective arrangements are in place and where they are necessary, that they work alongside off-site emergency plans prepared by the local authorities under COMAH or civil contingencies legislation, and in order to aid this process
- Provide local authorities with relevant information form the site that would facilitate effective action, including the arrangements established to help with the emergency response on site
- Provide information to the public including details of the dangerous substances, the possible major accidents and their consequences, and what to do in the event of an accident. This information should be easily understood by the public

From the above, it can be seen that both the European and national controls are rigorous in their information demands and requirements, for demonstrating effective control through a systematic understanding and management of hazards and associated risks. These controls have been continuously updated in the light of experience, often learned as a consequence of one or more major incidents. So while these controls evolve 'continuously' major incidents continue to occur. This will, in part, be due to the fact that those controls described above relate to Europe only, and they are not globally applicable, and even when controls may be in place, the level of enforcement may not be as rigorous as would be required for effective management of these high-risk activities.

This lack of rigour in enforcement of legislative controls may have played a part in a more recent incident in Tianjin, China, where on 12 August 2015, a series of large explosions rocked the port at Tianjin (BBC 2015). Some blasts were visible from orbiting weather satellites, while many of the firefighters that were sent to deal with the original fire were killed in the blasts. A total of 173 people were killed, and many more injured as a result of the fire and subsequent explosions that occurred at Ruihai International Logistics warehouses that had been used for the storage of amongst other materials, hazardous and flammable chemicals, including calcium carbide, sodium cyanide, potassium nitrate, ammonium nitrate and sodium nitrate. Approximately 700 tonnes of sodium cyanide was stored on site, and after the event, cyanide levels in water courses were found to be more than 20 times higher than recognised safe levels.

The investigators concluded that the double blasts were caused by spontaneous combustion of nitro-cotton in a container that had self-ignited under high temperature. The flame further ignited other chemicals, including the explosive ammonium nitrate, which then triggered the huge explosions. There have also been suggestions that water sprayed to douse the fire could have contributed to the blasts as a result of the generation of the highly explosive acetylene, which upon explosion could then have detonated other chemicals to create the large blasts that occurred. Given the devastation, it is unlikely that the precise cause will be known; however, beyond the mechanics of the explosions themselves, it seems to be widely accepted that poor regard for health and safety, together with the lack of regulatory control and enforcement were significant contributory factors; these blasts are a reminder of the absolute need for effective policy, regulation and enforcement in the management of hazardous chemicals.

The Tianjin incident was caused by goods/materials, but similar high-risk waste-related incidents are not uncommon, although they have not occurred on such a scale. For example, a fire at a waste management facility in Gloucestershire in the UK (Box 7.5) apparently started in the lab smalls area of the transfer station (HSE 2001). The site was classed as a 'lower tier' site under the COMAH (1999) Regulations and the incident constituted 'a major incident' under these regulations, and it was consequently reported to the EU.

Box 7.5 Hazardous waste-related incident within the UK

Although the precise cause of the fire is unknown, it is thought that the storage in close proximity (within the same drum) of incompatible wastes triggered a fire at a hazardous waste treatment site in Gloucestershire (Environment Agency 2013; Hitchings 2003). Laboratory smalls were stored next to Intermediate Bulk Containers (IBCs) filled with isopropyl alcohol

(IPA), and these were close enough for a pool of burning liquid to spread under the IBCs causing further spread of the fire to other flammable liquids stored nearby. When the spreading pool of burning IPA reached these containers, they are believed to have ruptured under the intense heat and exploded, producing large fireballs. The fire spread and caused the release of a range of chemicals, including hydrogen chloride gas and phosgene. The site in question was located close to the River Severn, which flooded very soon after the site fire, further enabling release of hazardous chemicals to the wider environment.

After the event, a series of management failings were identified that included (Environment Agency 2007):

- Unknown wastes on site and wastes that were not listed in the site inventory
- Lack of effective wastes segregation and lack of suitable fire breaks
- 'Rejected' wastes remaining on site

Incidents such as those described above highlight the need for especially effective control over hazardous chemicals—as products/goods or as wastes, an area that has seen significant changes in the last 50 years, and that has, for example, been reflected in the evolution of the Seveso Directive. While the incident in Gloucestershire, as outlined above, did not result in legislative changes, it did result in significant changes in the way that the relevant regulatory body in the UK (The Environment Agency) regulated such waste management facilities thereafter; soon after the event, the Environment Agency, together with the Health and Safety Executive (HSE) undertook a joint exercise looking at standards within the hazardous and chemical waste treatment sector. The findings of the audit included (Hitchings 2003):

- On a significant minority of sites, there was no comprehensive preventative maintenance programme
- On a minority of sites, labelling of wastes was not sufficiently adequate to ensure effective wastes segregation
- One site was found to be holding COMAH qualifying inventories but was not notified under COMAH

As a consequence of the above, Environment Agency inspection and control of such waste facilities changed from many, relatively short inspections to fewer, more detailed audits. In this way, although relevant controls were in place (e.g. COMAH; waste management licensing), the incident changed the enforcement activities in a way that provided for more effective control going forward. Evidence from the Tianjin Incident (above) has suggested that local enforcement was weak, and is a good example of the importance not just of having effective regulations in place, but of their diligent and rigorous enforcement and implementation of both national and wider controls. Such implementation not only helps to manage risks to the environment, but also helps to ensure that a so-called level playing field exists whereby everyone can compete for trade against a common cost baseline.

However, even when national and international controls are rigorously implemented, human error and lack of effective management control can still result in major incidents, and the explosion and fire at Buncefield in Hertfordshire, UK, in December 2005 provides a reminder of the importance of such elements. In this case, lack of effective management control, together with poorly understood and faulty level monitoring/control valves, resulted in enormous devastation. While over 40 people were injured (fortunately there were no fatalities), the ensuing fire, 'the largest seen in peacetime UK' (HSE 2011), engulfed over 20 fuel tanks and burnt for several days.

Very large volumes of fire-fighting materials were used to extinguish the fire at Buncefield, and some of the liquids (including perfluorooctane sulphonate from the foam, and hydrocarbons such as benzene and xylene) found their way into the adjacent environment, including a major aquifer from which potable water is extracted via, in some cases, unrecognised soakaways. According to the HSE (2011), 'The bunding at Buncefield had many flaws, which caused large volumes of fuel, foam and firefighting water to leak out of the bunds. Bunds were not impermeable and not fire resistant. The bunding was unable to handle the large volumes of firewater involved in the incident', and this was a major factor in the resultant pollution of the environment.

The Buncefield explosion 'was therefore further evidence that the major hazard industries had still not taken on board vital lessons' (HSE 2011).

Directives such as the 'Seveso Directives' and national controls to implement their requirements, are vital in helping to assure safe manufacturing environments, and in the twenty-first century, the effective use of information technology can provide important support.

By example, to support the implementation of Seveso III, the European Commission has invested in advanced Information and Communication (ICT) technologies including an entirely new reporting system, known as 'eSPIRS' (Box 7.6) and a Major Accident Reporting System known as eMARS (Box 7.7).

Box 7.6 eSPIRS

eSPIRS (EC 2016) provides information on establishments that are considered to present major hazards due to the potential accident risk associated with the presence of dangerous substances, as defined by the Seveso III Directive. There are more than 10,000 establishments reported on the eSPIRS database, made up of companies from 28 EU Member States, including Iceland, Norway and Switzerland. The purpose of the database is to support the Member States and the Commission in their risk management-related decision-making processes, by giving an insight into the geographical distribution of risk from Seveso establishments.

Box 7.7 eMARS (EC 2012b)

The main purposes of the online eMARS are to:

1. Raise awareness about potential failures leading to major accidents in Seveso establishments and the chemical industries in general
2. Facilitate the exchange of information on accidents and near misses that have occurred in Seveso establishments in Europe
3. Promote lessons learned on a global basis between the EU and its partner countries associated with OECD, UNECE, and UNEP as well as other countries of the world, where chemical accident prevention has gained attention
4. Provide the general public with access to accident information to aid local and national efforts to reduce chemical accident risks

These EU-funded databases, together with relevant EU Directives and supporting national regulations, work together to try to ensure that high-risk industries can continue to develop and grow in a way in which the risks are understood by all stakeholders and where appropriate control measures are in place to avoid major incidents in the future.

From the above, we can see that many major incidents have influenced the development of controls on high-risk activities; yet despite this, major incidents continue to occur. Such incidents often result from human error and/or lack of care and management, though in the next section we move on to look at controls on high-risk activities that have resulted from human concern over the global and local impact of chemicals and processes on the environment, and on human and animal health.

7.3 Issues of Global and Local Environmental Concern

On 27 September 1962, Rachel Carson published her book *Silent Spring* that documented the detrimental effects on the environment of the indiscriminate use of certain pesticides, and raised the profile of concerns relating to a group of chemicals known as Persistent Organic Pollutants (POPs). Her research looked, amongst other areas, at the use of dichlorodiphenyltrichloroethane (DDT) as a pesticide and its unintended consequences for wildlife, including birds. The book's title reflected concerns about the future health of the planet and the 'silent spring' without birdsong.

Since the publication of Silent Spring, it has been recognised that exposure to POPs can lead to serious health effects, including certain cancers, birth defects, dysfunctional immune and reproductive systems, greater susceptibility to disease and damages to the central and peripheral nervous systems (Secretariat of the Stockholm Convention 2009).

The use of POPs has been ubiquitous; in the mid-twentieth century, DDT was used extensively for a range of control measures including the eradication of insect vectors for malaria and typhus, and was very successful in this regard. However, the USA later banned the use of DDT, and this is cited by scientists as a major factor in the comeback of the bald eagle and the peregrine falcon from near-extinction within the USA (e.g. Stockstad 2007).

When dealing with materials such as POPs, the potential for transboundary movement renders national and even regional controls relatively ineffective, and therefore more globally effective controls are required if environmental and human health protection is to be assured.

In relation to POPs, this has been achieved through the Stockholm Convention and supported within Europe by the United Nations Economic Commission for Europe (UNECE), and especially the so-called POP Protocol (The UNECE Convention on Long-Range Transboundary Air Pollution (CLRTAP) on POPs), which entered into force on 23 October 2003 (United Nations, n.d.), and related regulatory controls.

The Stockholm Convention on Persistent Organic Pollutants (Secretariat of the Stockholm Convention 2009) was adopted on 22 May 2001 in Stockholm, Sweden, and entered into force on 17 May 2004.

The text of the Treaty (Secretariat of the Stockholm Convention 2009) acknowledges that:

- 'Persistent organic pollutants possess toxic properties, resist degradation, bioaccumulate and are transported, through air, water and migratory species, across international boundaries and deposited far from their place of release, where they accumulate in terrestrial and aquatic ecosystems'.
- They cause 'health concerns, especially in developing countries, resulting from local exposure to persistent organic pollutants, in particular impacts upon women and, through them, upon future generations'.
- 'The Arctic ecosystems and indigenous communities are particularly at risk because of the biomagnification of persistent organic pollutants and that contamination of their traditional foods is a public health issue'.

Given the above, it is clear that national and even regional controls would not be able to manage the impact of POPs, and that these international controls are vital for the protection of the environment and human health.

Between them, the UNECE POP Protocol and Stockholm Convention lay down the following control measures (list not exhaustive):

- Prohibition, or severe restriction of the production and use of intentionally produced POPs—specified in updated lists
- Restrictions on export and import of the intentionally produced POPs
- Provisions on the safe handling of stockpiles
- Provisions on the environmentally sound disposal of wastes containing POPs
- Provisions on the reduction of emissions of unintentionally produced POPs (e.g. dioxins and furans)

The Convention, together with the POP Protocol, shows how (for signatory countries) global concerns can be managed through more local/national actions. However, for such conventions to be effective, all countries that can significantly impact the environment through the use and management of POPs must implement and enforce the controls.

In 2004, the European Community signed both international instruments on POPs and adopted their requirements through relevant EC 'Decisions', while Regulation (EC) No 850/2004 of 29 April 2004 (EC 2004) complements earlier Community legislation on POPs and aligns it with the provisions of the international agreements on POPs. Because these controls are implemented through a European Regulation, the requirements are binding on all Member States, and therefore directly applicable within the UK.

As a result of their implementation, knowledge about the presence and use of POPs has facilitated more effective control and has been supported through controls to prevent the manufacture and use of specified chemicals, such as those on the disposal of PCBs (Polychlorinated biphenyls) (Box 7.8).

Box 7.8 Polychlorinated biphenyls

Polychlorinated biphenyls (PCBs) have excellent dielectric properties, longevity, non-flammability and resistance to thermal and chemical degradation, and for this reason, they were manufactured for use in electrical equipment, heat exchangers, hydraulic systems and several other specialised applications. However, because they were identified as being carcinogenic, their manufacture and use was banned in the late 1970s/early 1980s, according to national controls (e.g. National Archives n.d.-c).

Within Europe, controls on the disposal of PCBs were introduced through Directive 96/59/EC (EC 1996c) on the disposal of PCBs and polychlorinated terphenyls (PCTs), replacing Directive 76/403/EEC of 6 April 1976, and which was introduced with the aim of destroying PCBs, and equipment containing PCBs as soon as possible. The Directive required Member States to compile an inventory of equipment containing PCBs, and to develop a plan for disposal of the listed equipment. Those disposing of PCBs were required to keep registers of the quantity, origin, nature and PCB content of used PCBs delivered to them and to report this information to the competent authorities. The effect of these controls was to ensure that any equipment containing PCBs at a concentration at or above 0.05% was destroyed, or the PCBs replaced with less hazardous insulating fluids, and the segregated PCBs destroyed, thus removing any potential for future pollution of the environment or harm to human health from this particular type of POP. The list of substances to be 'eliminated' under the Convention are continuously updated, together with related exemptions, that for example, continue to allow the use of certain pesticides for malaria control, when banned for all other uses.

It is interesting to note that article 7 of Directive 96/59/EC on the disposal of PCBs and PCTs specifically prohibits the incineration of PCBs on ships (Box 7.9). This recognises that until well into the 1970s it had been common practice to incinerate PCBs and other chlorinated compounds at sea on one of three ships, the Vulcanus I and Vulcanus II and the Vesta (USEPA 1987). These ships operated outside territorial waters, and therefore were free from emission controls and could offer an inexpensive disposal service compared to on-land alternatives. However, even this was preferable to previous management options, in which it had been normal practice, well into the 1970s, for organochlorine wastes to be dumped into the North Sea and the Gulf of Mexico, and where in the North Sea, dumping reached a figure of approximately 90,000 tonnes per year (Compaan 1988). Such dumping at sea activities were eventually banned under the so-called Oslo Convention (Box 7.10).

Box 7.9 Incineration of waste at sea

The incineration of waste at sea began in 1969 in the North Sea (Compaan 1988) and was designed specifically for liquid organochlorine wastes with a maximum chlorine content of 70% w/w. In the USA, the use of ocean incineration was first proposed in 1974, when Shell Chemical Co. sought permission to use the Dutch-owned vessel Vulcanus I to incinerate liquid organochlorine wastes. In the years to come, tens of thousands of tonnes of organochlorine wastes, including PCBs, were incinerated in this way. However, as it became more widespread, opposition increased until its use was banned.

Box 7.10 The Oslo Convention

The Convention for the Prevention of Marine Pollution by Dumping from Ships and Aircraft (The Oslo Convention) was designed to control the dumping of harmful substances from ships and aircraft at sea (Center for International Earth Science Information Network n.d.). It was adopted on 15 February 1972 and came into force on 7 April 1974. The UK became a signatory to the Convention in 1975. The Convention prohibited the dumping in specified waters of a wide range of chemicals including halocarbons, mercury and mercury compounds, and other persistent materials, while the dumping of a further range of chemicals including arsenic, lead, copper, zinc and cyanides was restricted through permits. It was replaced by the Convention for the Protection of the Marine Environment of the North-East Atlantic (OSPAR Convention) on 22 September 1992, and subsequently amended on a number of occasions (OSPAR Commission 1992).

It is difficult to believe that significantly less than 50 years ago it was common practice to dump very large amounts of high-risk, highly toxic chemicals into our oceans, and only serves to demonstrate the need for regulatory control and effective enforcement of legislation to control these high-risk chemicals and activities.

The development of legislation relating to high-risk chemicals can be seen in the restriction of certain dangerous chemicals within electrical equipment, where within the UK for example, The Restriction of the Use of Certain Hazardous Substances in Electrical and Electronic Equipment Regulations 2012 (SI 2012 No.3032) (as amended), ('the RoHS Regulations 2012') implements the provisions of the European Parliament and Council Directive on the Restrictions of the use of certain Hazardous Substances in electrical and electronic equipment 2011/65/EU ('RoHS 2').

The Directive, implemented through these national regulations, with certain specific exemptions, has restricted the placing on the market of new electrical and electronic equipment (EEE) containing more than the permitted levels of lead, cadmium, mercury, hexavalent chromium and both polybrominated biphenyls (PBBs) and polybrominated diphenyl ether (PBDE) flame retardants in amounts exceeding the established maximum concentration values. When compared to the dumping of organochlorines in our oceans, these Regulations help to demonstrate just how far environmental control of high-risk chemicals and activities has come, and also provide a good example of the more recent approach to regulation in support of the circular economy, whereby limiting by legislation, the concentration of hazardous chemicals within new products, both limits potential for environmental harm when they become wastes, and also facilitates their re-use or recycling in secondary products. Such re-use and recycling of these high-risk/hazardous materials/chemicals not only reduce direct impacts, but will ultimately help to reduce the secondary environment and human health impacts associated with mining and processing activities of minerals and chemicals.

The Oslo Convention was borne out of a concern for the impact of the rather indiscriminate dumping of hazardous waste at sea; such restrictions caused those holding such materials to seek out the next lowest cost option (incineration at sea), and when this was ultimately stopped, progress to more sustainable options for the management of high-risk materials was adopted. Therefore, although this chapter began by highlighting nineteenth-century controls on high-risk activities, the 1960s and 1970s can be viewed as the nadir for environmental controls, and that to rectify this, a more coherent and sustainable policy approach has had to be adopted, and this policy approach will be explored within the next section of this chapter.

7.4 A Policy Approach

The previous two sections have demonstrated the influence and outcomes of specific incidents and 'environmental concerns' on the development, structure and implementation of regulatory controls on high-risk activities. The third major influence is that of environmental policy, and this can be influenced by a number of factors, including those covered in the previous sections. Since the time of the first environment legislation within the UK (such as the Alkali Act of 1863), and until relatively recent times, regulatory control of high-risk activities has been largely driven by the 'incident' and 'local concern' factors, and it has only been since entry into Europe that the policy approach has become more influential.

Early European Controls

Across Europe, environmental policy has had a significant influence upon the control of higher risk activities. Much of this policy has been implemented through Directives, therefore allowing for interpretation and associated variation at a national level, although the REACH regulations (EC 2006) provide one of a small number of exceptions to this 'rule'.

The framework for EU policy is identified within Environment Action Programmes (EAPs), each one being themed according to priorities and policy of the time; in relation to higher risk activities, early EAPs were very much about managing existing problems, such as those associated with high levels of oil pollution in the 1970s, while a later policy was able to take a more integrated approach to satisfy both business and environment needs. Policy objectives within the first and second EAPs (which covered the periods 1973–1976 and 1977–1981, respectively) resulted in Directives to better manage problems associated with, for example, waste oils and wastes from the production of titanium dioxide, as these were seen to be causing significant environmental risks across Europe. This provides an example of the cross-over between policy-driven controls of high-risk activities and those introduced in response to specific environmental concerns.

Although these early EAPs addressed certain environmental concerns, it should be recognised that the original Treaty of Rome did not provide for the adoption of legislation on environmental matters, and therefore these early EAPs and related legislation were introduced where implementation was necessary to support the operation of the 'common market'.

The Waste Oils Directive (EEC 1975a) provides an example, whereby in the pre-amble to the Directive, the need for control recognises that 'any disparity between the provisions on the disposal of waste oils already applicable or in preparation in the Member States may create unequal conditions of competition and thus directly affect the functioning of the common market'. It was the introduction of the Single European Act on 1 July 1987 that allowed for the introduction of legislation on environmental grounds, and indeed that introduced a number of related articles. The first of these articles set out the principles for EC environmental legislation as 'Action by the Community relating to the environment shall be based on the principles that preventive action should be taken, that environmental damage should as a priority be rectified at source, and that the polluter should pay'. These principles continue to form the cornerstone of European policy on the environment today, including that relating to high-risk activities, where these principles are as important to high-risk activities as those of less environmental and human health significance.

However, although early EAPs were not based on environmental aims, they nevertheless sought to protect health, safety and the consumer and, so, for example, in relation to wastes from titanium dioxide manufacture (EEC 1978b), the preamble to the Directive justified its introduction in recognising that 'whereas for waste from the titanium dioxide industry it is advisable to lay down a special system which will ensure that human health and the environment are protected against the harmful effects caused by the uncontrolled discharge, dumping or tipping of such waste'.

Concerns relating to titanium dioxide spanned both the first and second EAPs. The original 1978 Directive (EEC 1978b) recognised that while there were generic controls on the permitting of wastes and waste facilities within the original so-called Waste Framework Directive (EEC 1975b), more specific controls were needed to deal with the very particular hazards and risks associated with wastes from titanium dioxide production. This was to be done through 'a system of prior authorization as regards the discharge, dumping, storage, tipping or injecting of waste' while there was also to be monitoring of the wastes, and the environments into which they would be disposed. Member States had until 1 July 1980, to draw up programmes 'for the progressive reduction of pollution caused by such waste with a view to its elimination' and must 'fix the general reduction targets to be attained by 1 July 1987 at the latest and indicate the measures to be taken for each establishment'. The requirements of such Directives were, therefore, quite prescriptive in terms of requirements for implementation, but also very focused on dealing with issues at hand. These controls constituted a reactive approach to policy-making, reflective of the early stages of the European Union.

Although early EAPs were not based on environmental aims, they nevertheless sought to protect health, safety and the consumer and therefore, in relation to waste oils, not only was the then current level of pollution a concern, but there were also concerns related to the increasing levels of waste oil production, especially oil emulsions. As a consequence, a stated aim of the Waste Oils Directive was that the disposal of waste oils should have as one of their essential objectives the protection of the environment against the harmful effects caused by the discharge, deposit or treatment of these oils. It also sought to 'regulate the treatment, discharge, deposit and collection of waste oils and provide for a system of permits for undertakings which dispose of such oils, for compulsory collection and/or disposal of such oils in certain cases and for suitable inspection procedures'. In this regard, the UK was already ahead of the game, having introduced the need for licences (permits by another name) for the management of wastes through the Control of Pollution Act in 1974.

Within the UK, the implementation of controls on oil has been done separately by the various devolved assemblies. As identified above, many of the Directive requirements were met by early controls on wastes generally, such as those under Control of Pollution Act 1974, but it was not until 2001 that the first oil-specific regulations were introduced through the Control of Pollution (Oil Storage) Regulations 2001. In the absence of regulatory drivers, awareness advice on oils management had been available through campaigns, such as the Oil Care Campaign and through targeted initiatives by the regulators, but as shown in Fig. 7.1, a sustainable reduction in oil pollution incidents was only achieved after the introduction of specific regulatory controls in 2001.

The data in Fig. 7.1 clearly show a decline in reported oil pollution incidents since the introduction of the oil storage regulations in 2001, and while other factors, including various oil-awareness campaigns, may have influenced the decline, they provide an indication of the potential impact of such regulatory control over these and other higher risk activities. Yet, were it not for the presence of the Waste Oils Directive in the first instance, the drive for these national controls would be less significant.

In this way, early EAPs identified high-risk pollutants of particular concern, and through related Directives, introduced controls to manage

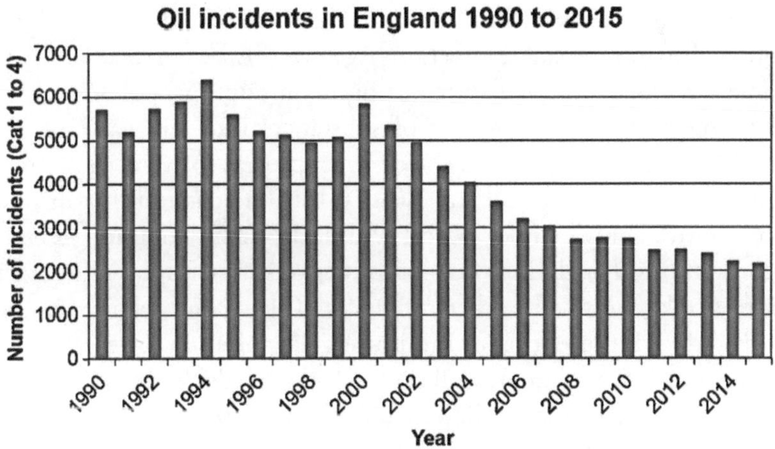

Fig. 7.1 Oil pollution incidents within the UK. © Environment Agency

the associated environment and human health risks; Member States would then have to introduce national controls to make these requirements binding locally.

By the time of the third and fourth EAPs (1982–1986 and 1987–1992 respectively), it had been recognised that simply controlling emissions to one medium may increase contamination in another, and therefore in the third EAP, the first elements of a 'sustainable development' approach to pollution control were evident.

Within the UK, this recognition of the importance of controlling emissions to all media formed the cornerstone of Part I of the Environmental Protection Act 1990, concerning the newly developed concept of Integrated Pollution Control (IPC). These new controls also recognised for the first time the need for specific controls on higher risk activities, and these were specifically identified within a Schedule to the related Regulations.

Part 1 of the Environmental Protection Act 1990 sought to control the most polluting processes and less polluting processes under separate control regimes, so that control was exercised in a way that was proportionate to the risk. These were Integrated Pollution Control (IPC) regime and Local Air Pollution Control (LAPC) regime. Those industries with the greatest potential to discharge polluting substances into air, land and water were subject to IPC and were regulated as Part A processes by the Environment Agency in England and Wales (at that time), and by the Scottish Environmental Protection Agency (SEPA) in Scotland. This regime was concerned with preventing or minimising pollution of the environment due to the release of substances into air, water or land. Certain less polluting industries (or their processes) were regulated by local authorities in England and Wales (SEPA in Scotland) for air pollution control only, under Local Authority Air Pollution Control Authorisations (LAAPC). Local authorities were also responsible for control of smoke, noise and other nuisances. This second regime was only concerned with preventing or minimising air pollution. Central to both regimes was the requirement that the 'Best Available Techniques Not Entailing Excessive Costs (BATNEEC)' should be used to prevent or minimise pollution.

The term 'Best Available Techniques' was defined to mean 'the most effective and advanced stage in the development of activities and their methods of operation which indicate the practical suitability of particular techniques for providing in principle the basis for emission limit values designed to prevent and, where that is not practicable, generally to reduce emissions and the impact on the environment as a whole' (EC 1996b).

This more integrated approach to pollution control was supported through the principle of Best Practicable Environmental Option (BPEO), in which BATNEEC was effectively used as the tool to demonstrate BPEO. As identified at the start of this chapter, the concept of BPEO had been introduced to replace BPM, that had been used for similar purposes since the nineteenth century.

To operate a process controlled under IPC a company had to obtain an authorisation (IPC or LAAPC), and this was the company's 'licence to operate', and when applying for an authorisation, the applicant had to demonstrate that the process used represented BATNEEC in seeking to prevent, minimise or render harmless polluting releases. BATNEEC covered all aspects of the way in which a process was operated, including design, emissions and supporting infrastructure, and for this reason the word 'technique' rather than 'technology' was used in BATNEEC.

This system of IPC control, which originated within the UK, was adopted by the European Commission who took the fundamentals of IPC and after modification, released the EC Directive 96/61 on Integrated Pollution Prevention and Control (EC 1996b) which, like IPC, required a range of industrial installations to be regulated by a system of integrated pollution control (i.e. a system in which emissions into air, water and land, plus other environmental effects, are considered together, and conditions are set so as to achieve a high level of protection for the environment as a whole). Permit conditions had to be based on the use of the 'Best Available Techniques' (essentially the same as 'Best Available Techniques Not Entailing Excessive Cost' in Part I of the 1990 Act, but where the 'NEEC' element was implied).

The IPPC regime followed the same principles as those under Integrated Pollution Control; however, 'pollution' was described more broadly (e.g. to include noise, heat and vibration), while energy consumption, waste generation and post-closure issues were also covered. IPPC also introduced new sectors for control including landfill, intensive farming and food and drink sectors, which had previously fallen under less restrictive controls.

The requirements of the IPPC Directive were implemented within the UK through The Pollution Prevention and Control Act of 1999 and Regulations made under this Act. This paved the way for a single pollution control regime for England and Wales, and was intended to help industrial operators move towards greater environmental sustainability.

Since its introduction in 1990 and subsequent modification by the EU in 1996, the principle of controlling higher risk activities and their impacts to all media has been implemented in different ways across the UK, where, for example, Scotland continues to use a waste management licensing regime for lower risk activities, and PPC controls for higher risk activities, whereas within England and Wales, the Environmental Permitting Regulations of 2007 (and subsequent Regulations) consolidated activities under waste management licensing and those under PPC permitting under a single 'environmental permit', but where the higher risk activities were still subject to more rigorous controls, were listed within Schedule I to the Regulations.

To support the identification of BAT, The European Commission's IPPC Bureau has developed and continues to develop, Best Available Technique (BAT) reference documents (otherwise known as BREFs) for all categories of industrial activity listed in Annex I to the IPPC Directive and each industry category has its own technical working group. These identify technologies and supporting procedures to ensure that BAT control is achieved for these more high-risk activities.

7.5 Sustainability Considerations for High-Risk Activities

As European environment policy has progressed from a reactive approach to one based more upon principles of sustainable development and the concept of the 'circular economy, so have the aims of European controls on high-risk activities developed accordingly.

Using waste oils as an example, the original Directive of 1975 was amended in 1987 by Directive 87/101/EEC (EEC 1987). As identified above, the original Directive sought to ensure safe and effective collection and management of waste oils, but this new Directive took this further by adopting a more recognisably 'sustainable development' approach by introducing the requirement for waste oils to be regenerated rather than burned for energy generation, where technical economic and organisational constraints allowed. The amended waste oils Directive also required that any oils that were recovered were free from hazardous chemicals as defined in the Directive on toxic and dangerous waste (current at that time) as well as being free from PCBs and PCTs above 50 ppm. Given the intention for oil regeneration, the need for these exclusions was clear.

This type of approach is more typical of current policy and related controls (e.g. see discussion of RoHS Regulations above) whereby products and goods are designed in a way that reduces the need for high-risk, hazardous components and that facilitates re-use and recycling in a way that supports the circular economy, and that helps to manage Earth and its resources in a more sustainable way.

The principle of sustainable development has been a cornerstone of European policy since the 5th EAP (1992–2000) which took a more holistic approach to managing the environment generally, and within the 5th EAP (CEC 1992), the use of the wastes management hierarchy as a tool to achieve sustainable wastes management was as applicable to high-risk, hazardous materials as it was to low-risk chemicals and processes. The logic of the wastes management hierarchy dictates that the greatest benefit to the environment and to sustainability, generally, can be achieved by designing out hazardous chemicals from products and/or using them at such low concentrations that their presence does not inhibit re-use and recycling activities.

In relation to higher risk activities, Chap. 6 the 5th EAP (CEC 1992) addressed the 'Management of Risks and Accidents'. This chapter recognised that the global production of organic chemicals alone had 'jumped from 7 million tonnes in 1950 to 63 million tonnes in 1970 and to over 250 million tonnes at present' and that 'Most chemical agents are potentially hazardous if incorrectly applied or if released in large quantities either by design or accident'. As a consequence, a process for dealing with existing chemicals was formulated in which 2000 high production chemicals would be identified, and after initial data collection, a preliminary assessment would be undertaken (1993 onwards). On the basis of this work, 200 priority chemicals would be identified, and a detailed assessment (25 chemicals per year for 8 years) would be undertaken, whereafter 50 priority chemicals would be subject to comprehensive risk reduction programmes between 1994 and 2000. Given the above timescales, it is a measure of the direct relationship between policy and regulation at the European level, that, in order to commence this process, Council Regulation on the evaluation and control of the risks of existing substances was published on 23 March 1993 (EEC 1993)!

All prior discussion has focused on regulatory control of high-risk activities, yet another significant driver of risk reduction in the management of high-risk activities is that of cost savings. When hazardous chemicals or materials are used in manufacture, raw material costs and process controls are invariably more expensive, and when the manufactured goods become waste, then disposal again tends to be significantly more expensive than the disposal of lower risk materials. Therefore, if hazardous, high-risk components can be designed out at the manufacturing stage, significant cost savings can accrue. Such cost savings are almost also inevitably associated with reduction in risks to the environment and human health.

While not all high-risk materials can be designed out of goods and materials manufacture, there are clear benefits to the environment when this can be achieved. As the concept of the circular economy gathers momentum, as it undoubtedly should, then opportunities for designing out high-risk chemicals will increase with potential knock-on effects (decreased demand) on high-risk industrial processes.

The information in Boxes 7.11 and 7.12 provide examples of the sort of activities that can reduce the impact of high-risk processes and chemicals through the use of waste minimisation and recycling programmes, rather than so-called end of pipe clean-up and treatment (e.g. emissions abatement techniques).

Box 7.11 Hazardous waste review leads to cost savings

A small engineering company reduced the volume of hazardous waste production by 1,00,000 litres per year by installing an oil filtration unit to separate oil/water mixes. The oil could be recycled, and the water used for cleaning operations. In addition to cost savings of £5500 per year from reduced hazardous waste disposal costs, the company also contributed to environmental benefits in reduced water demand for cleaning operations and reduced pollution potential from the storage of waste oils, pending collection (Envirowise 2005).

The information presented in Box 7.11 shows that significant environment savings can be achieved even within small companies through simple measures, such as low-cost in-house materials recycling with subsequent re-use of the recovered solvent, while information provided in Box 7.12 shows that process optimisation for optimum efficiency can also result in significant savings for the environment, while also saving money. Indeed, most measures introduced to save money will have a commensurate benefit to the environment.

Box 7.12 Solvent-reduction programme leads to cost and environmental savings for truck cab manufacturer

By implementing measures for more efficient spray techniques (e.g. switching to electrostatic spray guns) and in-house solvent recycling ABT products Ltd. were able to achieve:

- Net cost savings of £16,800/year
- Reduction in solvent waste of 75%
- Payback period on the combined measures of about nine months Envirowise (1998)

Globally, there are hundreds, if not thousands of case studies, similar to those identified in Boxes 7.11 and 7.12 relating to cost and environment savings from the more sustainable management of high-risk materials, especially in the design and management over the lifetime of a product. To date, such actions have been largely driven by interest in cost savings, but as identified above, such actions almost always tend to reduce potential environmental impacts at the same time. In a global market with variable legal constraints from nation to nation, it is difficult to see anything other than cost savings continuing as the key driver of such activities.

7.6 Summary

The development of UK and European controls has been examined through consideration of three key drivers, namely incident response, global and more locally identified environmental concerns, and environmental policy and strategy.

The picture that has developed has shown that the UK was the first to introduce legal controls on high-risk activities, and that such controls have been around since the nineteenth century. These early controls were introduced to manage 'local' pollution issues, and this 'reactive' approach has prevailed until relatively recent times. However, significant increases in industrial activity in the twentieth century, and the associated high level of the use and disposal of high-risk chemicals such as PCBs and other chlorinated organic compounds, has resulted in a more pro-active approach to the management of hazardous materials. This pro-active approach has been driven by European policy and related measures to support the sustainable management of the environment, but where international agreements, such as the Stockholm and Basle Conventions, have also had a significant influence.

At the same time, human health and environment disasters such as those at Seveso (Italy), Bhopal (India) and Buncefield (UK) have highlighted weaknesses in regulatory control and have initiated direct (reactive) improvements to European and National regulations and their implementation, to supplement this increasingly pro-active approach to the management of high-risk activities.

At a business level, manufacturers and users of high-risk chemicals have also looked to a more sustainable approach, in which high-risk chemicals have been substituted by less harmful alternatives, with associated reductions in both cost (in many but not all cases) and environmental impacts. As the principles of the 'Circular Economy' are further implemented through future measures and controls, this more sustainable approach to the management of higher risk activities and materials will become increasingly popular and important to the safe and effective management of Earth, its resources and its peoples.

References

Barthelemy, F, H. Hornus, and J. Roussot, et al. 2001. *Report of the General Inspectorate for the Environment 24 October 2001 Affair no. IGE/01/034 Accident on the 21st of September 2001 at a Factory Belonging to the Grande Paroisse Company in Toulouse Produced Jointly with the Explosives Inspectorate and with Help from the INERIS*. http://www.hse.gov.uk/landuseplanning/toulouse.pdf. Accessed 30 Jan 2017.

BBC. 2015. *China Explosions: What We Know About What Happened in Tianjin*. http://www.bbc.co.uk/news/world-asia-china-33844084. Accessed 24 Jan 2017.

CEC. 1992. *Towards Sustainability: A European Community Programme of Policy and Action in relation to the Environment and Sustainable Development*. COM (92) 23 Final, Vol. II, Brussels, 27 March 1992.

Center for International Earth Science Information Network. n.d.. http://sedac.ciesin.org/entri/texts/marine.pollution.dumping.ships.aircraft.1972.html. Accessed 23 Jan 2017.

Compaan, H. 1988. Waste Incineration at Sea. In *Pollution of the North Sea*, ed. W. Salomons et al. Heidelberg; Berlin: Springer-Verlag.

EC. 1996a. Council Directive 96/82/EC of 9 December 1996 on the Control of Major-Accident Hazards Involving Dangerous Substances.

———. 1996b. Council Directive 96/61/EC of 24 September 1996 Concerning Integrated Pollution Prevention and Control. *Official Journal* L257, 10/10/1996, pp. 0026–0040.

———. 1996c. Council Directive 96/59/EC of 16 September 1996 on the Disposal of Polychlorinated Biphenyls and Polychlorinated Terphenyls (PCB/PCT).

———. 2004. Regulation (EC) No 850/2004 of the European Parliament and of the Council of 29 April 2004 on Persistent Organic Pollutants and Amending Directive 79/117/EEC.

———. 2006. Regulation (EC) No 1907/2006 of the European Parliament and of the Council of 18 December 2006 concerning the Registration, Evaluation,

Authorisation and Restriction of Chemicals (REACH), Establishing a European Chemicals Agency, Amending Directive 1999/45/EC and Repealing Council Regulation (EEC) No 793/93 and Commission Regulation (EC) No 1488/94 as Well as Council Directive 76/769/EEC and Commission Directives 91/155/EEC, 93/67/EEC, 93/105/EC and 2000/21/EC.

———. 2008. Regulation (EC) No 1272/2008 of the European Parliament and of the Council of 16 December 2008 on Classification, Labelling and Packaging of Substances and Mixtures, Amending and Repealing Directives 67/548/EEC and 1999/45/EC, and Amending Regulation (EC) No 1907/2006.

———. 2012a. Directive 2012/18/EU of the European Parliament and of the Council of 4 July 2012 on the Control of Major-Accident Hazards Involving Dangerous Substances, Amending and Subsequently Repealing Council Directive 96/82/EC.

———. 2012b. *Welcome to the Major Accident Reporting System*. https://emars.jrc.ec.europa.eu/. Accessed 26 Jan 2017.

———. 2016. *eSPIRS*. https://minerva.jrc.ec.europa.eu/en/espirs/content. Accessed 26 Jan 2017.

EEC. 1975a. Council Directive of 16 June 1975 on the Disposal of Waste Oils (75/439/EEC).

———. 1975b Council Directive of 15 July 1975 on Waste (75/442/EEC).

———. 1978a. Council Directive of 20 March 1978 on Toxic and Dangerous Waste (78/319/EEC).

———. 1978b. Council Directive of 20 February 1978 on Waste from the Titanium Dioxide Industry (78/176/EEC).

———. 1982. Council Directive of 24 June 1982 on the Major-Accident Hazards of Certain Industrial Activities (82/501/EEC).

———. 1987. Council Directive of 22 December 1986 Amending Directive 75/439/EEC on the Disposal of Waste Oils (87/101/EEC).

———. 1993. Council Regulation No 793/93 of 23 March 1993 on the Evaluation and Control of the Risks of Existing Substances.

Environment Agency. 2007. *Review of Incidents at Hazardous Waste Management Facilities*. (Ref GEHO1107BNKU-E-P) November 2007 Version 1, pp. 13–20. Author.

———. 2013. *Review of Incidents at Hazardous Waste Management Facilities* (Ref LIT 6387), October 2013 Version 2.7, pp. 16–20. Author.

Envirowise. 1998. *GG128 Solvent Management in Practice: Industry Examples*, pp. 7–8. Author.

———. 2005. CS 498 *Hazardous Waste Review Steers Automotive Company Towards Cost Savings*. Author.

French Ministry of the Environment. 2008. Sheet Update No 5620 DPPR/PR/SEI/BARPI. *Seveso Accident: Release of Dioxins into the Atmosphere in a Chemical Plant.* http://www.aria.developpement-durable.gouv.fr/wp-content/files_mf/FD_5620_meda_Seveso_1976_ang.pdf

Hansard. 1972. HC Deb 25 February 1972 vol. 831 cc1667–72. *Dumping of Cyanide (Nuneaton).* http://hansard.millbanksystems.com/commons/1972/feb/25/dumping-of-cyanide-nuneaton. Accessed 13 Jan 2017.

Hitchings, A. 2003. Major Incidents at Waste Treatment Sites—Case Studies and Lessons. IChemE Symposium series no. 149, pp. 131–141.

HSE. 2001. *Report for the Deputy Prime Minister the Right Hon John Prescott MP into the Major Fire on 30 October 2000 at Cleansing Service Group ltd Sandhurst.* Joint Report by The Health & Safety Executive and the Environment Agency, p. 10. Author.

HSE: COMAH. 2011. *Buncefield: Why Did It Happen? The Underlying Causes of the Explosion and Fire at the Buncefield Oil Storage Depot, Hemel Hempstead, Hertfordshire on 11 December 2005.* Author.

OSPAR Commission. 1992. *OSPAR Convention of 1992, as Amended.* http://www.ospar.org/site/assets/files/1290/ospar_convention_e_updated_text_in_2007_no_revs.pdf. Accessed 26 Jan 2017.

Secretariat of the Stockholm Convention. 2009. *Stockholm Convention on Persistent Organic Pollutants (POPs) as amended in 2009. Text and Annexes.* http://chm.pops.int/TheConvention/Overview/TextoftheConvention/tabid/2232/Default.aspx. Accessed 26 Jan 2017.

Stokstad, E. 2007. Species Conservation. Can the Bald Eagle Still Soar After It Is Delisted? *Science* 316 (5832): 1689–1690.

The National Archives. n.d.-a *The Control of Major Accident Hazards Regulations 1999.* http://www.legislation.gov.uk/uksi/1999/743/contents/made. Accessed 24 Jan 2017.

———. n.d.-b *The Control of Major Accident Hazards Regulations 2015.* http://www.legislation.gov.uk/uksi/2015/483/contents/made. Accessed 25 Jan 2017.

———. n.d.-c *The Environmental Protection (Disposal of Polychlorinated Biphenyls and Other Dangerous Substances) (England and Wales) Regulations 2000.* http://www.legislation.gov.uk/uksi/2000/1043/contents. Accessed 23 Jan 2017.

United Nations n.d. *United Nations Treaty Collection.* https://treaties.un.org/Pages/ViewDetails.aspx?src=TREATY&mtdsg_no=XXVII-1-g&chapter=27&clang=_en. Accessed 26 Jan 2017.

USEPA. 1987. *Ocean Incineration: Background and Status. National Service Center for Environmental Publications (NSCEP).* Prepared by Margarete S Steinhauer and Christine E Werme, September 4 1987. https://nepis.epa.gov/Exe/ZyNET.exe/9100OOPF.txt?ZyActionD=ZyDocument&Client=EP

A&Index=1986%20Thru%201990&Docs=&Query=&Time=&EndTime=&SearchMethod=1&TocRestrict=n&Toc=&TocEntry=&QField=&QFieldYear=&QFieldMonth=&QFieldDay=&UseQField=&IntQFieldOp=0&ExtQFieldOp=0&XmlQuery=&File=D%3A%5CZYFILES%5CINDEX%20DATA%5C86THRU90%5CTXT%5C00000026%5C9100OOPF.txt&User=ANONYMOUS&Password=anonymous&SortMethod=h%7C-&MaximumDocuments=1&FuzzyDegree=0&ImageQuality=r75g8/r75g8/x150y150g16/i425&Display=hpfr&DefSeekPage=x&SearchBack=ZyActionL&Back=ZyActionS&BackDesc=Results%20page&MaximumPages=1&ZyEntry=1&slide

Van der Velden, P.G., C.J. Yzermans, and L. Grievink. 2009. *Mental Health and Disasters*. Chapter: 28, Cambridge University Press, pp. 473–496. https://www.researchgate.net/publication/232169565_Enschede_Fireworks_Disaster

Varma, Roli, and Daya R. Varma. 2005. The Bhopal Disaster of 1984. *Bulletin of Science, Technology & Society* 25 (1): 37–45.

Vogel, D. 1986. *National Styles of Regulation: Environmental Policy in Great Britain and the United States*. Ithaca, NY: Cornell University Press.

Ken Westlake Bsc (Hons) MPhil Phd MCIWM CEnv has 30 years' experience in waste management consultancy and training, undertaking work in a number of overseas countries as well as in the UK. While at the Environmental Safety Centre at Harwell laboratory, he managed several long-term projects for the (then) Departments of Energy and Environment. He has managed research and consultancy activities for The World Bank, the European Union, the UK Department for International Development (DfID), The British Council and The Environment Agency. Within the UK, Ken provides consultancy support in relation to resource efficiency/circular economy, and this has included support to the Envirowise programme and to the East Midlands Improve Your Resource Efficiency programme. He has published more than 40 peer-reviewed articles and has written and/or contributed chapters to several books on wastes management.

8

Environmental Risk Management and Assurance

Duncan Giddens

8.1 Securing and Assuring Compliance—A Starting Point for Risk Management

The Purpose and Evolution of Environmental Regulation

In today's fast changing world, a successful organisation needs to be agile, adaptable, innovative, collaborative and integrated. The same is true of environmental 'regulation' and 'management'. This chapter explores how integrated approaches and interventions can help organisations to make better decisions to manage risks and opportunities associated with the environment. In doing so, it considers how environmental regulation should evolve from its past emphasis on state-based 'instruction'—organisations being told what to do, to self-governance—organisations being accountable, informed and competent to manage their interactions with the environment.

D. Giddens (✉)
Optimus Management Solutions Ltd, Somerset, UK

J. Foreman (ed.), *Developments in Environmental Regulation*, Palgrave Studies in Environmental Policy and Regulation, DOI 10.1007/978-3-319-61937-8_8

The purpose of environmental regulation has been to protect people and wildlife against the impacts of industrial and other potentially harmful activities. To date, the emphasis and main requirement of regulation has been for organisations to comply with state-imposed rules or regulations. This has been treated by many (but not all) organisations as a bolt on to business-as-usual activity and the environment may not have been recognised as a core part of the business. As a result, the state has had to intervene, by way of environmental regulation, to address this 'market failure'.

Environmental regulation has also been used to increase energy and resource efficiency including by recycling and reuse of materials defined as 'waste'. Process efficiency and waste minimisation have been important commercial drivers, but there has also been a realisation by public and private sectors that a linear extraction-consumption-disposal approach is not sustainable and that a circular economic approach is the way forward.

A third imperative is a growing recognition that all parts of the natural environment are valuable resources. This has led to the concepts of Natural Capital and Ecosystem Services, which are explored in Sect. 8.3 of this chapter.

Many organisations have come to realise that the best way to manage their interaction with the environment—including compliance with laws and other obligations—is as an integral part of business risk management throughout their value chain. This effectively brings the management of environmental risks and opportunities into the mainstream rather than the periphery of economic activity. This perspective has been reflected in the revision (in 2015) of ISO 14001, the international standard for environmental management systems (EMSs), one of the most widely used tools for improving the environmental performance of organisations.[1]

Box 8.1 sets out concepts and approaches that are explored throughout this chapter.

Box 8.1 Key concepts and approaches

Business performance management	Natural capital and ecosystem services
Compliance and compliance management	Regulation and regulations
Dependencies and dependency modelling	Risks and opportunities
Innovative solutions	Standards and standardisation
Integrated approaches	Sustainable resource management (circular economy)
Management systems	Value chain optimisation

Environmental Risk

Environmental risk can be described as the potential for change, or level of uncertainty, within the environment. Risks and changes can be adverse—have a negative impact—or beneficial—have a positive impact.

The term environmental aspect is also used to describe how an organisation interacts with the environment, and again, aspects can have a positive or negative impact on the environment. Aspects can be broadly seen as risks and opportunities.

Adverse environmental risks pose a hazard, or impact, with negative consequences to the environment, for example, an industrial process that releases substances into the environment at levels that are toxic to wildlife.

The term 'significant impacts' is often applied to risks which are hazardous enough to require positive intervention to prevent or mitigate adverse effect(s) on the environment.

In addition to risks to the environment from (human) activities, there are also risks to humans and wildlife that arise from a changing environment, for example, floods, storms and drought. Risk to and from the environment are often interrelated—as is the case with human-induced climate change. An increase in flooding events may be (in part) caused by increased levels of greenhouse gases in the atmosphere emitted from industrial processes. Or drought might be caused by higher levels of water abstraction and use.

Environmental risks and impacts may be short term in nature where the effect is of a short duration, transitory and does not persist. These could be viewed as acute. Others may act over a long period and be persistent—these could be viewed as chronic. Both short- and long-term impacts can have a significant effect on the environment. But the measures and resources needed to manage short- and long-term impacts may be very different.

The precautionary principle is a key approach used in environmental regulation in which measures are taken to prevent something harmful occurring in the future, provided it can be foreseen, is likely to happen and the consequences would have a significant impact. For example, limiting the accumulation of hazardous substances to prevent a major accident or installing a bund to contain potential spillages.

In managing environmental risks, organisations should consider the uncertainties associated with their understanding of the way the environment behaves and the uncertainties in knowing what are good and bad human actions for the short and long term.

Environmental Opportunities

Opportunities can arise from managing risks to and from the environment. The foremost opportunity may be the ability to enhance the environment in terms of its resilience, diversity, quality and quantity. This will help achieve a better quality of life for humans and wildlife and may bring economic benefits, for example, by ensuring environmental resources/capital/services are resilient and sustainable and therefore useable.

Another area of opportunity lies with the development of new technologies and techniques to protect and enhance the environment which can improve the efficiency of an organisation, or groups of organisations, and may present new business opportunities, for example, improvements in energy and resource efficiency.

The concept and application of 'Best Available Techniques' (BAT) is informed by this thinking. The intention of BAT is to define measures that provide the necessary level of protection to the environment while also being the most suitable (practicable, technically effective and cost effective) for target organisations to use within their business environment. There is an argument that the adoption of BAT for environmental protection provides opportunities for businesses to both reduce/optimise their impacts on the environment and develop more effective and efficient ways of working for the business as a whole, for example, by improving processes, reducing waste and energy consumption. See Chap. 6 for details of the opportunities that can be achieved through the implementation of BAT.

Why Do We Need Environmental Regulation?

Effective management of environmental risk should both protect the environment and make business sense. So why do we need environmental regulation? In most countries, the state and its authorities/agencies impose rules and standards on regulated organisations. The regulatory authorities typically check compliance with these rules and measures, usually via site inspections, and take enforcement action—when necessary—to ensure compliance. This is the traditional 'command and control' model of regulation.

At this point, it is worth considering the meaning of regulation. For many people, regulation means the setting and enforcing of rules. In the case of environmental regulation, most of the 'rules' applicable in the UK come from the European Union (post Brexit, this may change!). This is often called Regulation with a 'big R'.

However, in the broadest sense of the word, regulation (with a 'small r') means providing control over something to make sure it works as intended. For example, a watch has a regulatory mechanism to make sure it gives the right time. Or an engine is regulated so that it operates within safe limits and runs efficiently. Or a manufacturing process is regulated to make sure that products are made to the right quality.

Within many fields of protection, for example, environment or health and safety, unregulated markets tend to move towards the lowest level of protection. This is partly because the costs of protection have been externalised (or skewed) and are not borne by those impacting most on the area of protection, such as the environment. And partly because a proportion of organisations and individuals operate to maximise financial gain over and above any other consideration.

Environmental regulation is used to ensure a minimum level of protection of the environment (people, wildlife, habitats and resources) to meet society's current expectations. Within the broader concept of sustainable development, regulation is one approach that can be used to ensure that society lives within the planet's natural resource limits and resilience. But direct regulation alone is not the solution—a mix of approaches is needed to reflect the different circumstances and types of organisation including their drivers, values and interaction with the environment.

Environmental regulation should ensure that organisations whose activities impact on the environment (arguably, that is all organisations) manage and mitigate both the risks they pose to the environment and the risks from the environment to them. Regulatory intervention is also used to:

- Set a level playing field by applying minimum acceptable standards
- Prevent and mitigate accidents, deliberate acts, illegal activities
- Address specific failures that lead to poor environmental outcomes (e.g. poor air, water, soil quality)
- Assess and secure compliance with rules/regulations
- Impose sanctions for continued and serious non-compliance with the rules

Securing Compliance

Securing compliance with legal requirements has been the focus of environmental regulation and is an initial step, or starting point, for managing environmental risk.

This section considers three key questions:

- What is compliance?
- What should be considered when assessing and managing compliance?
- What are the most effective tools and techniques for securing compliance?

What Is Compliance?

A simple definition of compliance is 'meeting requirements in full'. A more specific definition of 'legal compliance' is 'conformity with the law in such a way that the intended outcome is realised' (IMPEL 2016, Consultation response to European Co-operation for Accreditation regarding the review of guidance document, EA-7/04).

The international standard for EMSs, ISO 14001: 2015, uses the term 'compliance obligations' and defines these as 'legal requirements and other requirements'. It also sets out that 'legal requirements are those an organisation has to comply with by law' and 'other requirements are those that an organisation has to or chooses to comply with'.

The standard makes a distinction between *mandatory requirements*—such as applicable laws and regulations—and *voluntary commitments*—such as organisational and industry standards, contractual relationships, codes of practice and agreements with community groups or non-governmental organisations.

Regulators often view compliance as meeting the requirements—or conditions—of a permit to operate (see Chap. 6).

It is very unlikely that an organisation will be able to achieve absolute (100%) compliance with all its obligations, all the time. It is therefore useful to consider different levels of compliance and which of these are acceptable. The Environment Agency (England) uses a compliance-enforcement model which sets out four levels of compliance with environmental permits. The upper two levels are considered acceptable, while the lower two levels are considered unacceptable.

What Should Be Considered When Assessing and Managing Compliance?

Three key considerations are useful when assessing compliance:

- *How significant or important is the rule/obligation/requirement?* What would be the consequence of not complying? It can be argued that resources should be targeted on the most important rules which—for the environment—when followed provide the greatest protection and mitigate the highest risks.

- *How compliant is an organisation at present and what is its direction of travel?* Most regulators focus resources on organisations falling into one of the bottom two levels, but there is a good argument for making sure that any organisation is moving upwards (i.e. it reviews progress and seeks to continuously improve). This is indicated by the red arrow in Fig. 8.1.
- *Who is responsible and accountable for achieving compliance?* If you accept that it is responsibility of the organisation subject to compliance to meet its obligations, it follows that the controlling and decision-making level of management is accountable for meeting and managing compliance obligations.

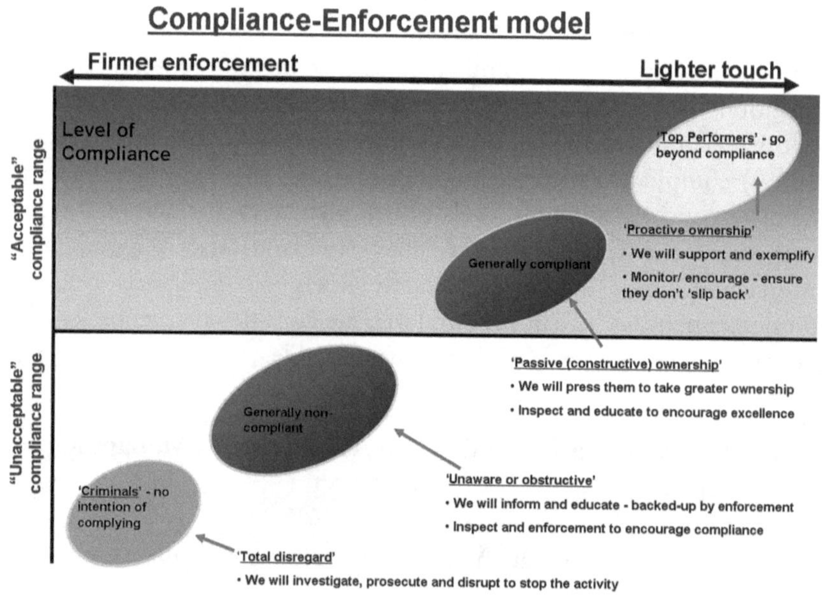

Fig. 8.1 Environment Agency Compliance—Enforcement Model (SNIFFER Report ER34 2013)

What Are the Tools and Techniques for Securing Compliance?

There are number of tools currently used to promote, assess and assure compliance. These include:

- Regulatory guidance—generic and sector specific (the role of guidance is covered in chapters on permitting and sector approaches)
- Risk appraisal
- Regulatory inspections and audits
- Internal audits (first party) and external audits (second and third party)
- Compliance assurance schemes
- Corporate reporting on performance
- Environmental and/or compliance management systems (CMS) and standards

Other tools used for enforcing compliance, including financial incentives, and penalties and sanctions are covered in other chapters of this book

The following sections explore the use and effectiveness of the tools used for promoting, assessing and assuring compliance.

Risk Appraisal

Many regulators assess the risks posed by regulated organisations' activities, including to the environment. This is necessary for a 'risk-based' approach to regulation!

The Environment Agency in England uses the operational risk appraisal (Opra) framework to assess the risk that an operator's activities pose to the environment (Environment Agency 2014). An Opra assessment considers four attributes: the type of activities carried out (complexity attribute), the location of the activities and sensitivity to pollution, types of polluting substances (emissions) and the performance of the operator in complying with permits and managing risks.

Opra is also used to help determine the level of annual cost recovery charges to operators based on the overall risk rating, or score, given to an operator.

Although Opra is generally viewed as a robust approach, recent reviews have highlighted shortcomings around its emphasis on 'fixed' attributes and lack of criteria and incentives for improving performance, including risk management.

In 2010, the Hampton Review of Progress (Environment Agency 2010) reported that 'as it is currently formulated, improved compliance does not always have a significant impact on a business's Opra score and therefore on its levels of inspection, or fees. There is therefore scope for Opra to offer a greater incentive for increased compliance, for example if there was a greater differential in fees and inspection levels between compliant and non-compliant businesses of a similar nature. We recognise, however, that incentives in charging schemes are difficult to apply where the principle of cost-recovery is rigidly applied'.

The Hampton Review also stated that 'the EA also indicated that it recognised its (Opra's) limitations in incentivising particular behaviour by charging and is considering other incentive mechanisms such as providing a choice of regulatory regimes when appropriate such as earned autonomy or self-certification. Businesses said they would very much welcome self-certification, understanding that this would place greater responsibility for compliance on them, with concomitant consequences if they mis-certified'.

Following on from Hampton, and subsequent internal EA reviews, it is clear that Opra, and similar risk appraisal approaches used by other environmental regulators, could be improved to better reflect the capacity and actual performance of an operator in managing environmental risks, rather than focusing on 'fixed attributes' such as location and emissions. A further progression could be to encourage a more standardised approach to risk management that is applicable to all activities carried out and which is an integral part of the management system(s) used by the regulated organisation. This could be done by including 'standard' environmental management criteria in the performance appraisal 'module' so that performance can be measured and compared across organisations and sectors. The Environment Agency has introduced standard rules permits based on sector risk assessments and a standard permit condition for a permit holders' management systems. These provide a good starting point for standardised, risk-based performance criteria.

Regulatory Inspections and Audits

Regulatory inspections are typically site visits carried out by regulatory officers which are used to check levels of compliance. Supporting information on compliance is provided by regulated organisations such as the results of emissions monitoring and evidence of duty of care.

Inspections may be planned/routine or random. They may include some elements of auditing but are essentially a snapshot check on the level of compliance at a point in time, usually restricted to permits and any directly imposed rules. The emphasis of most regulatory inspections is to identify 'non-compliances' and to bring these to the attention of the site operator.

Audits provide a more structured method of assessing compliance and how it is being managed. In addition to checking on the symptoms of compliance, they provide a means of checking how compliance is being managed and the root causes of poor performance.

To be effective, inspections—spot checks on compliance—must be made frequently enough to address risks and to provide a history of performance. With many thousands of regulated organisations, some with multiple permits across different sites, site-based inspections are very resource intensive for regulators. Also, the effectiveness of an inspection programme is largely based on deterrence—that is the chances of being caught and suffering a punishment—or sanction rather than education and incentives to encourage and reward better management practice.

Regulatory audits are an extension of inspection and, because they are used to check 'whole' permit compliance, may be even more resource intensive.

As regulatory resources are cut,[2] fewer and fewer inspections are being carried out by regulators. Since 2002, the number of waste licence site visits by the Environment Agency has been reduced from over 100,000 (Environment Agency 2008) to around 20,000 per year currently (2016).

As visits are targeted on those sites with the worst performance and with the highest risk and/or public interest (subject to most complaints), this means that there is less regulatory oversight for the majority of permitted organisations. However, this has not been matched by an increase in regulatory support for operator-based compliance assurance. The situation in the UK *is highlighted by a report in ENDS (26 July 2016) that* '40% of English and Welsh councils and port health authorities are not respecting government guidance on inspections of industrial installations and mobile plant. The situation is worst for the most environmentally at risk activities. The most frequent explanations offered for this were lack of staff, departmental restructuring, plant closure, poor record keeping and mobile plant being out of the area'.

The Hampton Implementation Review of the Environment Agency (Environment Agency 2008*) concluded that* '*there is a move towards fewer inspections but there is a lack of evidence about the effectiveness of these interventions in achieving regulatory outcomes or reducing burdens*'.

The Hampton Review of Progress in 2010 reported that waste management and the pollution prevention and control regime both showed a 40% reduction in levels of inspection since 2007.

The following sections explore how assurance-based interventions can support and, ultimately, replace site inspections to provide more risk-based and tailored environmental regulation.

Internal and External Audits

All organisations are managed in some way to ensure that they achieve intended results. At the most basic level, this management will cover the core activities needed to run the organisation (e.g. financial, legal and people management) and activities to make/deliver products or services. The more complex the organisation, the more structured management is needed to make sure all elements work as intended. Audits are a way of checking performance against requirements set by management and can be applied to all the activities and disciplines within an organisation.

Audits provide a structured method of evaluating performance with a recognised and repeatable methodology. They can be used to check all or specific parts of a system or standard (or permit). They can be used to perform sample checks that build up to a whole system check.

Internal audits may be used independently or as part of a wider management system approach. Typically, these audits are carried out by a part of the organisation not directly involved or responsible for the activities being checked. This ensures a degree of impartiality and independence.

The further removed those carrying out audits are from the 'line of accountability' for the audited process or activity, the more independent the audit is likely to be. Ultimately, audits can be carried out by another organisation altogether. These are called external audits and may be carried out by a second or third party. A second party may have a link with the audited organisation (e.g. a sector trade association) and may not itself be checked. A third party is completely independent and may be checked to ensure competence and impartiality. An example is an EMS Certification Body (CB) accredited by a National Accreditation Body, such as the UK Accreditation Service (UKAS). UKAS keeps a register of accredited bodies and the disciplines or standards they are accredited to inspect, audit or certify.

Of course, there is a cost to an organisation when it establishes an internal audit function or employs external auditors, or both. However, this cost is likely to be small compared to the costs of poor-quality products/services or the costs to remedy environmental damage caused by a serious pollution incident. There may also be a longer term cost arising from lost reputation and business, in addition to the more immediate costs of putting something right.

Environmental auditing within a framework of self-regulation should make good business practice, as well as being a complementary measure to regulatory inspection. And auditing is one component of a broader approach that can be called compliance assurance.

Compliance Assurance

Assurance can be defined as a set of activities carried out by an organisation that checks and reports on performance to interested parties.

Compliance assurance is one example of this approach, applied to compliance with the law relating to environmental activities.

Independent assurance may also be called verification. A good example of how assurance can be provided is by the use and verification of management system standards (MSS) such as ISO 14001 for environmental management or ISO 19600 for compliance management. While the organisation that uses the standard is responsible for meeting its compliance obligations, verification of how well this is being done can be carried out by independent third parties, for example, when auditing EMSs and standards.

There are also specific compliance assurance schemes, for example, those set out in the case studies below. Both schemes are 'owned' by the Environment Agency, and organisations take part on a voluntary basis, subject to meeting certain 'scheme entry' criteria. However, in both cases, the verification of compliance is carried out by independent third parties.

Box 8.2 Case Study 1, Environment Agency Pig and Poultry Assurance Scheme (Environment Agency 2016b)

This is a voluntary assurance scheme for intensive pig and poultry producers who are achieving a high standard of compliance with their environmental, that is, Environmental Permitting Regulations (EPR), permit.

CBs inspect member farms and collect data. Wherever possible, they do so when carrying out an audit for another scheme, such as Red Tractor Assurance or the Lion Code of Practice, to decrease the number of regulatory visits to farms.

Once a farm has joined the scheme the CB is given a copy of the permit and any variations or other information relevant to permit compliance. The CB collects information that helps to assess whether a farm is complying with its permit.

Farms in the scheme have an annual inspection. In a three-year period, one inspection is carried out by the Environment Agency, and in the other two years, the CB carries out inspections. Farms that are not in the scheme are inspected by the Environment Agency on a frequency determined by the environmental risk of the farm.

The CB will charge for carrying out their visit, but it is likely to be much less than the regulatory charge reduction. By reducing the number of visits, permit subsistence charges are lowered by 50% per year.

Box 8.3 Case Study 2, Monitoring Certification Scheme (MCERTS) (Environment Agency 2016a)

Business emissions to air, land and water are regulated under European and UK laws to protect the environment and human health. To comply with these laws, organisations need a permit to operate, which usually requires monitoring of emissions.

Emissions monitoring must meet quality requirements set by the Environment Agency including both continuous monitoring and periodic monitoring.

MCERTS is the Environment Agency's MCERTS. It provides the framework for businesses to meet emissions monitoring requirements, provides the regulator with confidence in monitoring of emissions to the environment and is used to help targeting and prioritising independent auditing of water and air monitoring.

MCERTS is used to approve instruments, people and laboratories. It also includes Operator Monitoring Assessment (OMA), alongside technical guidance and standards for different types of monitoring.

OMA applies to the monitoring of emissions to air and discharges to water from industrial processes regulated under the EPR. It used to strengthen the Environment Agency's assessment of operators' self-monitoring including monitoring undertaken on behalf of operators by contractors using a consistent and transparent approach.

The two case studies below illustrate how modern data management can help secure and report compliance with environmental requirements.

Box 8.4 Case Study 3, OpenSpace web portal (Viridor 2011)

OpenSpace is an interactive web portal developed by Viridor, with the Environment Agency, that enables users, for example, regulatory officers and, ultimately, other interested parties, to view environmental compliance data via a series of structured queries on any computer linked to the internet.

OpenSpace is the product of a four-year change programme within Viridor that has examined and changed every aspect of the way they schedule, obtain and process environmental data. Viridor and its partner organisations developed automated and standardised scheduling of environmental work across the UK waste and environmental industries. This has included a database scheduling link to a laboratory and use of Personal Digital Assistant (PDA) handheld devices to obtain field data in a consistent and quality-assured manner.

All information visible within OpenSpace is underpinned by multiple layers of quality assurance to ensure the key data and associated management actions can be communicated in a timely and contemporary manner to the regulator, the Environment Agency.

Box 8.5 Case Study 4, Emisoft and Statoil (referenced in *IEMA* Environmentalist Journal 2016)

In 1993, Emisoft were tasked with a joint project from Statoil and Hydro, the two largest oil and gas companies in Norway, to design a system to keep track of their environmental performance.

While Statoil has focused on compliance, improving sustainability performance and continuously enhancing the efficiency of environmental accounting and reporting, Emisoft has focused on developing a system that provides traceability and transparency in assuring compliance. At times, a new requirement from Statoil informed a development by Emisoft. At other times, an improvement of the solution such as a new configuration or a report could enable Statoil to work faster or better manage and understand their performance.

The Emisoft solution is used to manage Statoil's environmental performance throughout the organisation. Environmental data is managed in one common system. Data is accessed, updated and reviewed at site level and aggregated, monitored and reported on corporate level. Having access to quality data is fundamental for how Statoil conducts its business—a carbon intensity KPI is included in the company CEO's score card.

Environmental authorities introduced a goal for zero-discharge requirements for chemicals along the Norwegian continental shelf. Statoil, Hydro and ConocoPhillips challenged Emisoft to develop a system to support a zero-discharge strategy for chemical use. Emisoft developed a model for calculating chemical discharge to enable the companies to reach the goal. The zero-discharge goal was met in 2005, and chemical discharges have been consistently lowered in subsequent years.

The EU Emissions Trading Scheme (EU ETS) is another example of a complex regulatory requirement where Emisoft has provided a system that secures compliance. The EU ETS includes requirements for the submission of reports for all CO_2 emissions, third-party verification of the emission data and payment of CO_2 quotas. Statoil uses Emisoft to make all the necessary calculations and required reports, relying on the system's inbuilt traceability and transparency to ensure complete, accurate and valid results.

Statoil communicates its sustainability credentials on its website, providing statistics about energy needs and sustainability alongside detailed explanations of how they monitor their environmental impact, and their annual disclosure reports. This information is produced by the Emisoft management and reporting system.

Corporate Accounting and Reporting

Corporate accounting and reporting is a way of keeping track of performance and communicating an organisation's commitment and performance to interested parties. It can be a key tool for assuring environmental compliance and performance.

Corporate responsibility (CR) reporting, as well as many other elements of risk and performance management, has evolved to meet the needs of a changing world. Over the last 20 years, CR has developed from environment, health and safety reporting to sustainability reporting and now includes supply-chain issues, ethics, gender rights, and anti-slavery. Key factors driving growth in CR include regulation, consumer demand, stakeholder pressure, branding and public relations, director liability, as well as specific sector and local issues.

There is now a greater focus on reporting issues that significantly impact, or are material to, a business. The major reporting frameworks—the Global Reporting Initiative, the International Integrated Reporting Council and the Sustainability Accounting Standards Board—promote this push for greater 'materiality'.

In addition to being a driver for changing organisational behaviour, CR can be a driver for investor behaviour change. This is potentially far more powerful. Although investor behaviour generally changes more cautiously and slowly than organisational behaviour change, it is happening on carbon-related issues because of the increasing weight of scientific evidence, carbon reporting, political and NGO lobbying, and investor organisations with a specific carbon remit (ENDS Report, 25 July 2016).

A key aspect of CR which needs to improve is independent assurance of reports. Currently only 20% of reports have third-party assurance. Without such assurance, one can question whether reported performance can be taken on trust. Ideally, all good companies should have their financial and non-financial reports independently assured. The Non-Financial Reporting Directive is due to come into force in the UK (in 2017). This could be a driver to expand corporate reporting and help to bring more rigour to reports.

Stakeholders should be able to rely on the completeness and accuracy of reported qualitative information and quantitative data, together with the accuracy of the system used to put it together. This is important for issues such as environmental compliance if regulators are to take account of corporate reporting as part of assurance.

In addition to regulation benefitting from CR reporting, regulation is a driver for increased rates of reporting. And, as rates increase in the largest companies, it can be expected that reporting will become more common in small- and medium-sized companies within the supply and value chains of larger companies—in much the same way that the uptake of standards for quality and environmental management has been driven through supply chains.

Management consultants, KPMG, have carried out a series of surveys on CR, reporting from 1993. The 2015 report, Currents of Change (KPMG 2015), predicts that 'CR reporting becomes the norm, driven by regulation', and highlights the following trends:

- Almost three quarters of N100 companies now report on CR. The current rate of CR reporting among the G250 is over 90% *(N100 are the largest 100 companies in each of 45 countries across 16 industry sectors, totalling 4500; G250 are the world's largest companies by revenue, listed in the Global Fortune 500 in 2014).*
- Including CR data in annual financial reports is now a firmly established global trend. Almost three in five companies do this now, compared with only one in five in 2011. This is being driven by regulation in many countries. The eight countries with the highest rates of CR disclosure in financial reports all have legislation that requires it.
- There is a growing trend of regulations requiring companies to publish non-financial information. In the survey report, KPMG states: 'What will change the game is the introduction of more regulation requiring companies to report non-financial information. Non-financial reporting will become required business practice. Companies now need to focus on what they will report and how best to integrate their financial and non-financial information.'
- Third-party independent assurance of CR information is now firmly established as standard practice among the world's biggest companies (G250): almost two-thirds invest in assurance.

Environmental Management Systems

While the 'instruct and punish' mode of regulation is still suitable for a minority of serial offenders, or for those organisations with little or no management capability, the majority of legally operated organisations will benefit from a more consensual and partnership-based approach that educates, encourages and exemplifies good management practice. Many regulators are now either using or planning to use compliance approaches that make use of organisations' management systems. There are several reasons and drivers for moving in this direction:

- Many large and complex organisations already use management systems and a significant proportion (around 30%) employ independent (accredited, certified) auditors to verify their performance.
- The lack of resources available to environmental regulators to maintain historic levels of inspection, coupled with diminishing returns for inspection-based regulation and enforcement for organisations with effective management systems in place.
- A management system approach offers a structured and standardised method for managing compliance and for monitoring, evaluating and improving overall environmental performance.
- Proportionality—a fit for purpose management system can reflect the type, complexity and potential risks of activities carried out.

Since 2008, the Environment Agency has set a condition in all new or varied EPR permits to require the operator to implement a management system to ensure compliance with permit requirements. Demonstrating compliance with this condition is a pre-requisite for joining the Pig and Poultry Assurance Scheme, as described in Case Study 1, above. It is also likely to be a pre-requisite for any other assurance schemes.

Many European environmental regulators are also looking at how to use of accredited, certified EMSs for compliance assurance. The premise is that the private sector has established standards covering a wide range of requirements, for example, ISO 14001 for EMSs, ISO 50001 for energy management, ISO 31000 for risk management and ISO 19600 for compliance management.

At the same time, organisations are being driven or encouraged to consider their corporate risks and opportunities together. For example, all new and revised ISO MSS follow a common 'High-Level Structure' which helps to drive a more integrated approach to management. Later in this book, a case study is presented on MSS 1000:2014, a standard intended to enable organisations to create fully integrated management systems across all activities. The MSS has been created by a team of experts within the Chartered Quality Institute (CQI) Integrated Management Special Interest Group. (CQI Special Interest Group 2016).

The risks arising from areas previously treated separately in 'silos', for example, environmental protection, health and safety, food safety, quality assurance, are increasingly being considered in a wider context of business risks and opportunities and as a part of sustainable business practice. This has several advantages including:

- Considering environmental, social and economic risks and opportunities together
- The potential to optimise activities throughout the value chain, not just site-based operational aspects
- More integrated use of management systems and standards

When applied to securing compliance, the key elements of a systematic management approach may include (not an exhaustive list):

- *Top management commitment* to meeting compliance obligations
- *Identifying requirements, risks and opportunities* for compliance within the context of the organisation
- *Securing and maintaining* compliance, actively doing what is necessary to be compliant and resolving potential or actual non-compliances or breaches
- *Measuring, monitoring and evaluating* the level of compliance to determine compliance status.
- *Assuring and reporting* compliance and providing information to others on compliance status and record
- *Reviewing and improving the CMS* in light of changing circumstances

Management approaches are covered in more detail in Sect. 8.3, Innovative and Integrated Approaches.

8.2 Designing and Choosing Interventions to Achieve Better Outcomes

Introduction—Why Are Interventions Needed?

In a perfect world, all organisations would undertake and manage their activities in ways that do not have negative consequences for other people and organisations or for the wider environment in which they operate. The same may be said at the macro level for the operation of economic and financial markets. However, there are internal and external (commercial) pressures that result in industrial and/or economic activities that may have harmful impacts on the natural environment, or which may be dangerous/hazardous to people. So, there are a range of 'interventions' which are used to bring activities and markets within boundaries and rules set by society.

There is an established body of environmental legislation from international treaties, the European Union and UK national government/assemblies. This legislation is implemented via prescribed measures such as directly applied rules, permits to operate, and a range of other interventions available to public authorities. A good example is the European Industrial Emissions Directive (IED), which is covered in Chap. 6.

Interventions may also be created and used outside of state-based regulatory regimes. Some are supply chain-driven obligations that may give organisations little or no choice to comply if they wish to remain and operate within the supply chain.

Other 'softer' types of intervention are based more on influencing and changing behaviour and may be driven by stakeholders such as peers, local communities or customers.

And there are measures which organisations take themselves because it makes good business sense. The drivers may be financial, reputational, ethical or moral.

Interventions typically fall into 5 or 6 categories as shown in Box 8.1.

Box 8.6 Typology of interventions (Adapted from Defra Instrument Selection Guidance 2013)

- *Direct regulation (permits/licences, inspection/monitoring, enforcement/ penalties).* These have relatively certain outcomes but are resource intensive and may need to be targeted (rationed).
- *Economic instruments* (e.g. *taxes, trading schemes, financial incentives*). These have less certainty of outcome but greater flexibility for businesses to choose least costly options.
- *Information-based approaches* (e.g. *intelligence led approaches, campaigns*). Dependent on availability of data/information and ability/ resources to analyse and utilise.
- *Partnership/joint working* (e.g. *collaborative projects/initiatives*). Can encourage rapid action, flexible to changing circumstances.
- *Self-regulation* (e.g. *standards such as ISO 14001*): action motivated by financial, customer/supply-chain or reputational influence.
- *Support and capacity building* (e.g. *campaigns, advice and guidance*). In addition to supporting self-regulation, interventions may be aimed at 'interested parties', such as citizens, consumers, investors and NGOs who may be able to influence the behaviour of regulated organisations or provide information to regulatory agencies.

The Department of Environment, Farming and Rural Affairs has produced a guide to instrument (intervention) selection for policy-makers and regulators (Defra 2013). The guide helps users to think of ways to influence business behaviour to achieve environmental objectives. It provides:

- A summary of the main types of instrument available, and when they might be appropriate (20 different ways to influence business behaviour are described and compared)
- Questions to help users decide which options could work for the businesses being targeted
- Cross-cutting themes help users think through effective implementation
- Models to help analyse and compare options

How to Identify, Design and Choose the Right Interventions

Regulators and policy-makers need evidence to answer the question 'which approach works best, and in what circumstances?' The model approach set out in Fig. 8.2 provides a framework to consider the 'dependent factors' applicable in any given situation and the interventions that may be effective in securing a goal or set of goals.

Evidence may be available for some factors but much will be 'expert' judgement of the users and others. It is good practice to gather together

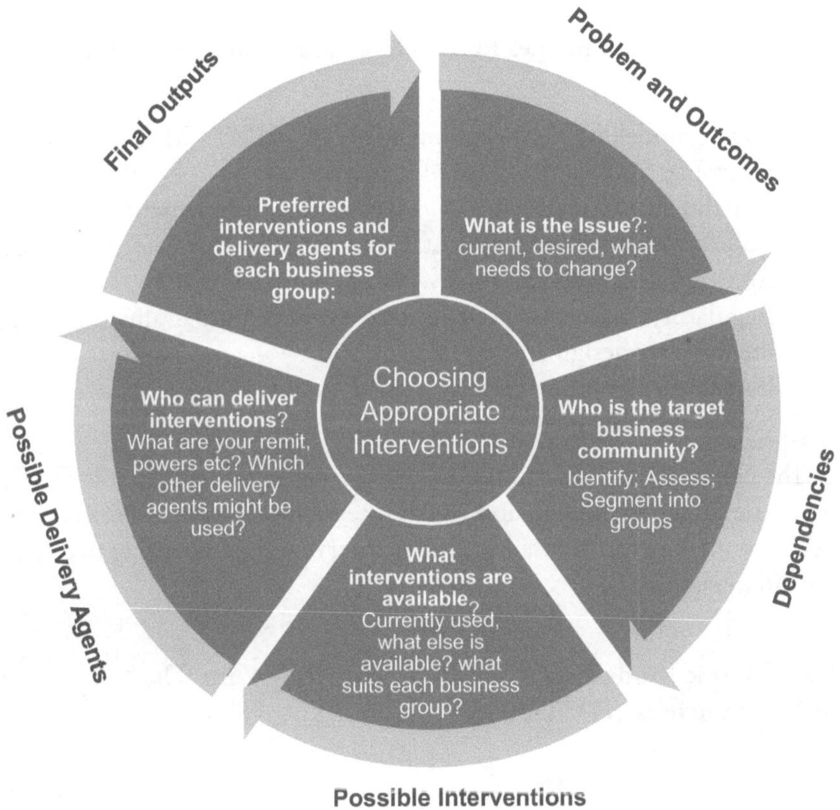

Fig. 8.2 Model for choosing interventions (IMPEL Report: 2014/12)

a group of experts to work together on this exercise. The dialogue and the exploration of what a programme, policy initiative, target community or sector looks like may be as valuable as actually building and running a model. Research, and researchers, in a range of disciplines—especially the economic and social sciences—can help provide the answers.

This model is based on the plan-do-check-act (PDCA) cycle which is the basis of most management system-based approaches, and many similar 'policy cycle' models used by government departments and bodies.

Techniques and Tools for Choosing Interventions

There are several techniques for choosing and evaluating interventions, including:

- Logic modelling as recommended by HMT Magenta Book, Guidance for Evaluation (HMT 2016)
- Dependency modelling as adapted for Implementation and Enforcement of Environmental Law (IMPEL) project, Choosing Appropriate Interventions (IMPEL Report: 2014/12)
- Regulatory impact assessments used to choose and evaluate government policy design and implementation
- Post-implementation reviews used to evaluate the success—or otherwise—of policies

The next two sections summarise (a) how logic modelling tools can be used to evaluate the design and use of interventions and (b) how dependency modelling can be used to choose interventions according to different circumstances.

Using Logic Modelling to Evaluate the Design and Use of Interventions (HMT 2016)

Logic modelling provides a framework for any type of programme by explicitly linking activities and processes with short- and long-term

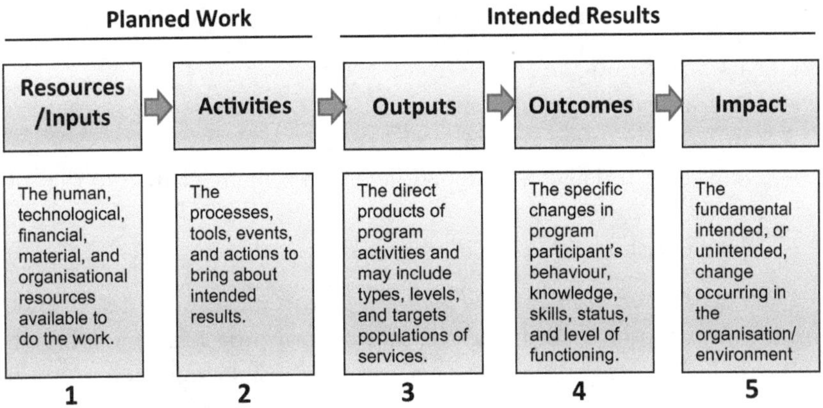

Fig. 8.3 Logic modelling

outcomes. It can help to visualise and predict how an intervention—or a set of interventions—is expected to work. It can be used to capture and describe the theory, assumptions and evidence used by those designing, applying and evaluating policies, goals or interventions (Fig. 8.3).

Using a logic model means following a chain of reasoning or '*if… then…*' statements that connect a programme's parts so that an organisation (or individual) can determine:

1. *if* you have access to the necessary resources/inputs,
2. *then* you can use them to accomplish your planned activities, and *if* you accomplish your planned activities,
3. *then* you will (hopefully) deliver the amount of product and/or service you intended, and *if* you accomplished your planned activities to the extent you intended,
4. *then* your participants will benefit in the ways you planned, and *if* these benefits to participants are achieved,
5. *then* the desired changes in systems and the organisation can be expected to occur.

Logic modelling can be used to help:

- Factor in timing and sequence of actions/activities
- Demonstrate a complex situation (set of variables) in a format that can be easily understood and visualised
- Determine linkages and dependencies which are critical to the success of the intervention
- Make decisions and prioritise resources

During *programme design and planning*, the model can be used to think through assumptions, and evaluate different approaches to solving the problem and meeting the need. This encourages stakeholders to examine best practice research and practitioner experience associated with each approach.

In *programme implementation*, the model forms the core of a focused management plan including identifying and collecting the data needed to monitor and improve the programme to accomplish short- and long-term goals. The model is a work in progress that can be revisited and revised as results become known.

For *programme evaluation and reporting*, the model presents programme information and progress towards goals in ways that inform, advocate for a particular approach, and teach stakeholders what works and doesn't work. Large-scale impacts most frequently occur sometime after the conclusion of the formal programme.

Finally, the logic model argues for an inclusive approach to all stages of the programme, involving all levels of stakeholders including management, implementers and recipient participants.

For policy-makers/implementers, the logic model approach can be used to evaluate:

- Whether the planned policy interventions result in expected outcomes/impacts?
- What can be done better/what can be learnt/what examples of good practice can be identified?
- What dependencies or assumptions are there and what are the critical activities and links on which success depends (see dependency modelling below)?

- What type of data is needed/what evidence is needed (*what decisions will be taken based on the evaluation findings)*?
- What metrics could be used to measure project or policy outcomes?

Using Dependency Modelling to Help Choose Interventions (IMPEL Report: 2015/18)

The process of choosing interventions to meet policy goals is problematic because:

- The success of interventions is dependent on different sets of circumstances
- The evidence available about each intervention is often limited and difficult to find
- The 'systems' in which regulators operate are often complex and multivariable

Dependency modelling can support decision-making within complex systems by setting out thinking in a way that helps the user to *consider* the dependencies to achieving their goal and how these fit together in a model. Once you have constructed and 'run' a model, it will also help identify which areas to focus attention on and which interventions might best address the dependencies which most threaten achievement of the goal.

Dependency models can help users compare the effectiveness of different interventions, such as earned recognition, voluntary approaches, fiscal incentives and trading schemes, as well as more traditional interventions such as advice and guidance, inspection and enforcement.

To be used effectively, dependency modelling users should be knowledgeable about the given situation, organisations and the interventions that can be applied. The quality of any decisions that you come to over interventions will be related to the logical structure of your dependency model, the accuracy of the data and assumptions used to set links between dependant factors—in other words, how well you have described and quantified the interrelationships between factors.

iDEPEND Dependency Modelling Tool

iDEPEND is a tool that has been developed to analyse what a successful outcome of a process depends on in any system, be it an organisation, endeavour, business, factory, machine, decision, plan or idea—anything in fact for which a suitable model can be built.

iDEPEND was tested and adapted for use by environmental practitioners as part of the IMPEL project 'Choosing Appropriate Interventions' (IMPEL Report: 2015/18). The project identified dependency modelling as a technique that could help policy-makers and practitioners choose and then apply interventions to achieve better environmental compliance.

iDEPEND uses a 'Top–Down' approach to analyse risk and allows users to create a framework of pre-requisite conditions—or dependencies—for achieving a goal and mitigating the risks, or uncertainties around the goal.

A goal can be anything from avoiding disasters to successfully making the right decision or launching a new enterprise. Once a 'goal' has been defined and a dependency model has been constructed, *iDEPEND* can help:

- Predict the probability of achieving goals
- Predict the risks to those goals
- Find those dependencies which are most pivotal to goals and on which you need to focus (the dependencies which are the most likely causes of success and failure)
- Thereby showing you where you might apply 'countermeasures' (or interventions)
- predict the effectiveness of those interventions on achieving your goals
- And thereby find ways to reduce risk and increase resilience in the 'system' you are trying to manage to achieve your goals

The model in Fig. 8.4 was developed by the author as part of an IMPEL project to determine the critical elements of a CMS (IMPEL Reports: 2011/04—2013/15—2014/16—2015/19—2016/19).

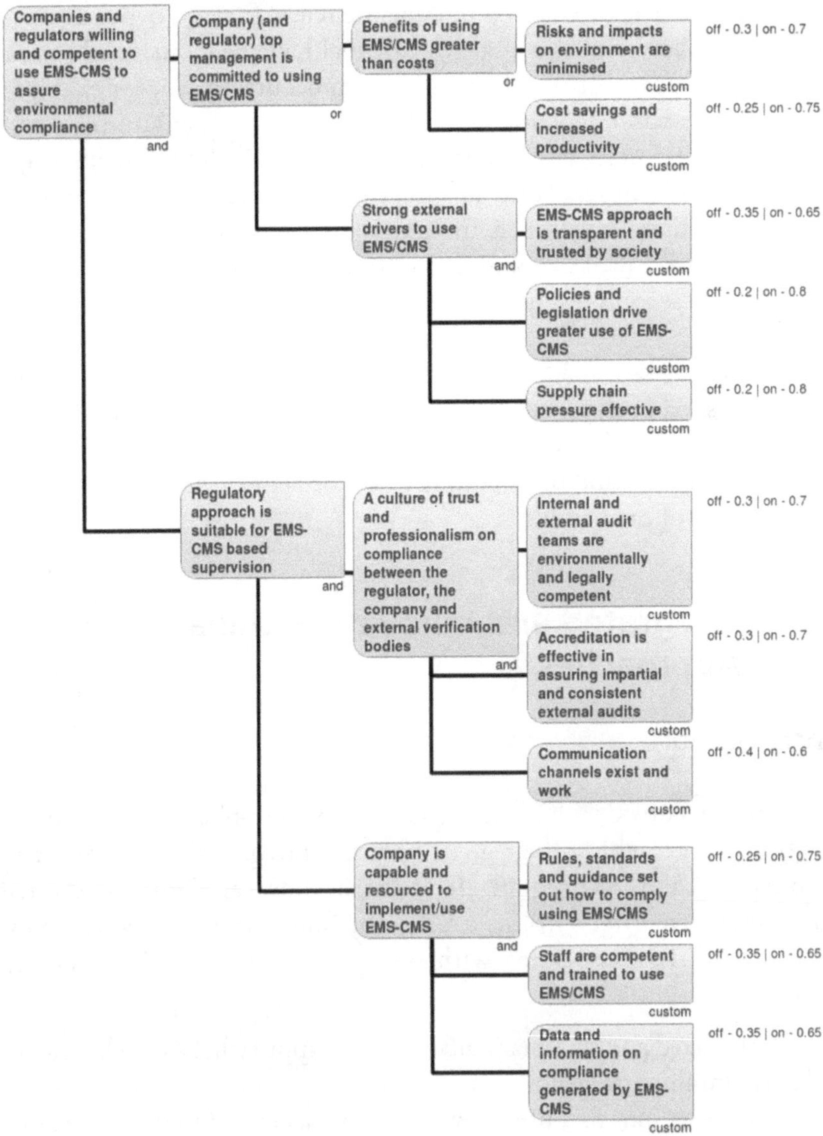

Fig. 8.4 iDEPEND model for an Environmental Compliance Management System (EMS-CMS)

This model has three levels of dependencies. The end dependencies, or 'leaves', that form the final (third) level have been assigned probabilities that represent the chances of them occurring. So, for example, the top right-hand dependency 'risks and impacts on environment are minimised' has been assigned a probability of 0.7 (70%) likelihood of being true. The model 'adds up' the end (leaf) probabilities and considers how these are linked to the goal ('companies and regulators willing to use EMS-CMS to assure environmental compliance') via the first and second levels to calculate the probability of achieving the goal.

Models may be iterated and refined to improve the chances of achieving the desired goal, and to determine the dependencies which are most critical for success. Iterations can be made to improve the structure and links of the model and by obtaining more, or better, evidence to increase the accuracy of predictions.

8.3 Integrated and Innovative Management Approaches

Introduction

This section considers how organisations can work better and be more successful in reaching their goals by using integrated and innovative approaches and techniques. It looks at how applying integrated and innovative interventions can lay the foundations for better management of risks associated with the environment and, in so doing, how:

- Integrated and systematic management approaches can help achieve environmental goals
- Organisations that use innovative approaches and tools can be more successful and sustainable
- Environmental regulation can (and should, the author would argue) be based on risk management and value chain optimisation

Integrated Regulatory and Management Approaches

Over the last 20 years, environmental policy-makers and implementation bodies have sought to integrate environmental legislation, practice and support systems.

The UK developed the Integrated Pollution Control regime to bring together protection of air, water and land environmental media. The EU further developed this approach by introducing the Integrated Pollution Prevention and Control (IPPC) Directive, and more recently the Industrial Emissions Directive (IED). These approaches have brought together under one regime, regimes that were previously formulated and implemented separately to protect air, water and land (largely waste) environmental media.

The EU has also developed Framework Directives for Water and Waste, which bring together legislation and standards for waste and water.

And the EU is developing a package based on the concept of a circular economy, which is intended to embed sustainable resource management requirements in EU legislation. A new British Standard, BS 8001, has been developed to set out how organisations can implement the principles of a 'circular economy'. This standard brings together a number of existing approaches such as management systems, process efficiency (also termed 'lean'), value chain optimisation, industrial symbiosis and natural capital/ecosystem services (not an exhaustive list).

The environment is a key area of EU legislation, based on the premise that it is more effective and efficient to manage environmental activities, risks and impacts that go across national boundaries in a coherent, integrated and framework approach, rather than in the traditional/historic media-based 'silos' and nation-state-based regimes. The European Environment Agency provides sound, independent information on the environment for those involved in developing, adopting, implementing and evaluating environmental policy, and also the general public. In close collaboration with the European Environmental Information and Observation Network (Eionet) and its 33 member countries, the EEA gathers data and produces assessments on a wide range of topics related to the environment.

This has come full circle in England, with IPPC and IED, EU directives as well as other permitting regimes, being implemented via the EPR to a common framework approach.

European Union Network for the Implementation and Enforcement of Environmental Law (IMPEL)

The IMPEL Network is an important driver for more consistent and joined up implementation of EU Environmental Law, bringing together practitioner experience and developing good practice guidance and innovative tools. The tools include:

- Doing the Right Things methodology (IMPEL 2006–2009)
- Supporting the implementation of Integrated Risk Assessment (IMPEL 2014)
- Choosing appropriate interventions including the use of dependency modelling (covered in Sect. 8.2, above)
- Compliance assurance through company compliance management systems (covered in Sect. 8.2, above)

Integrated Management System Approaches

The risks arising from areas such as environmental protection, health and safety, food safety and product/service quality are increasingly being considered in a wider context of business risks and opportunities and as a part of sustainable business practice. This has several advantages including:

- Considering environmental, social and economic risks and opportunities together
- A greater emphasis on value chain aspects, as well as fixed site, and operational aspects

The move towards more integrated approaches is happening in both the business/commercial world—broadly, the private sector—and in the public/state-led enterprises and authorities—broadly, the public sector.

In addition to potentially leading to better outcomes across the above areas, integrated approaches can help organisations to make better use of private- and public-sector resources. For example, it is more efficient and effective to audit an integrated management system that covers environment, occupational health and employee safety, rather than to audit several systems in isolation, at different times and to different management criteria.

An effective approach to management and innovation can be summarised under the three 'disciplines' below. Each discipline is illustrated with baseline, progression and best practice examples, where baseline is a minimum expectation, progression builds on the minimum and best practice is the expectation for class leaders. They can also be viewed as levels of maturity of management and management systems.

1. *Leadership and commitment to set direction/strategy and achieve results*, for example, to:

    - engage with interested parties and understand their expectations/requirements (baseline)
    - demonstrate compliance with legal and other requirements (baseline)
    - manage risks and opportunities (progression)
    - set your own objectives, as well as complying with rules set by others (best practice)

2. *The right people and skills to deliver results*, for example, with the ability and competences to:

    - operate safely and within required standards or limits (baseline)
    - identify and evaluate risks and opportunities and set objectives to manage/mitigate (baseline)
    - identify, evaluate and manage risks (progression)
    - understand how the organisation functions within its value chain (best practice)

3. *Management and control to monitor and evaluate success in achieving results*, for example, the ability and means to:

    - define, measure and assess performance (baseline)
    - review and improve performance (progression)

- manage interactions within the value chain and wider society (best practice)
- recognise, plan for and manage change in a complex world (best practice)

Other attributes that should be demonstrated by organisations building a culture of innovation and effective change management include:

- *removing barriers* to innovation, minimising risks and exploiting opportunities
- *influencing and working with others* in the supply and value chain, including contractors, partners and investors. Good examples are energy (carbon) efficiency within a sector/supply chain and resource efficiency through circular economic approaches within supply and value chains
- *considering whether the organisation is best at managing incremental, or step, changes* and the ability to decide which will give the desired results
- *using data, information and knowledge management to improve performance*, for example:
    - data sharing and open data initiatives (e.g. data portals, open data-sets, hackathons)
    - knowledge management (e.g. knowledge transfer networks, communities of practice)
    - public/community initiatives/crowd-sourcing
    - social media and smart apps (e.g. invasive species monitoring, litter/fly-tipping reporting)

Three Examples of Innovative and Integrated Approaches

This section sets out three examples of innovative approaches that could be used to regulate, or govern, interactions with the environment within an integrated management framework. Two (Baldrige and CQI) are generic management approaches which may be applied to any type of activity and are designed to do this in an integrated and consistent way. The approaches are:

1. Natural capital and ecosystem services including use of the Natural Capital Protocol
2. Baldrige Performance Excellence Framework
3. CQI Management System Specification and Guidance (MSS 1000:2014)

Example 1, Natural Capital and Ecosystem Services

Natural capital can be defined as the world's stocks of natural assets which include soil, air, geology, water and all living things. From natural capital, society receives benefits known as 'ecosystem services'. The Natural Capital Coalition has been established to promote and drive this thinking and to deliver practical applications, such as the Natural Capital Protocol, summarised below.

An ecosystems approach takes an integrated approach to the management of the 'benefits' that nature provides. It enables people and organisations to make clear links between a healthy natural environment and their social and economic well-being and involves all relevant parties in finding and implementing solutions to environmental problems.

Ecosystem services attempts to put an economic value (quantitative benefit) on living systems. This helps to move away from economic approaches that treat natural resources as both disposable and unlimited 'commodities' and treats them as valuable and limited resources that need to be managed on a sustainable basis.

The concepts of natural capital and ecosystem services are starting to inform environmental policy development, for example, the development of the 'circular economy' as the best way to be resource efficient and operate sustainably. This can be compared to natural cycles, for example, water and carbon, where there is a balance between use, state, retention and release of these substances within the environment (the planet). There cannot be a complete loss of such elements from a closed system, but they may be effectively 'locked in', and not available for use.

The Natural Capital Protocol (2016) is a standardised framework to help organisations identify, measure and value their impacts and dependencies on nature. It seeks to build on existing tools and techniques, such as management systems and corporate reporting standards (as described in Sect. 8.2 of this chapter) to identify measure and value natural capital and focuses on improving internal decision-making.

Baldrige Excellence Framework for Measuring and Improving Organisational Performance (Baldrige Foundation 2016)

The Baldrige Performance Excellence Program supports organisational performance excellence in the US and throughout the world. The programme was founded by the Baldrige Foundation, whose vision is:

> We will be recognized around the world as the premier foundation for the promotion of performance excellence in all sectors of the economy. We will inspire organizations to achieve excellence in all they do with pride and societal responsibility. Together with our stakeholders we will create value through research and deployment of validated, cutting-edge management practices. We will focus on the future and manage for agility, innovation and creativity.

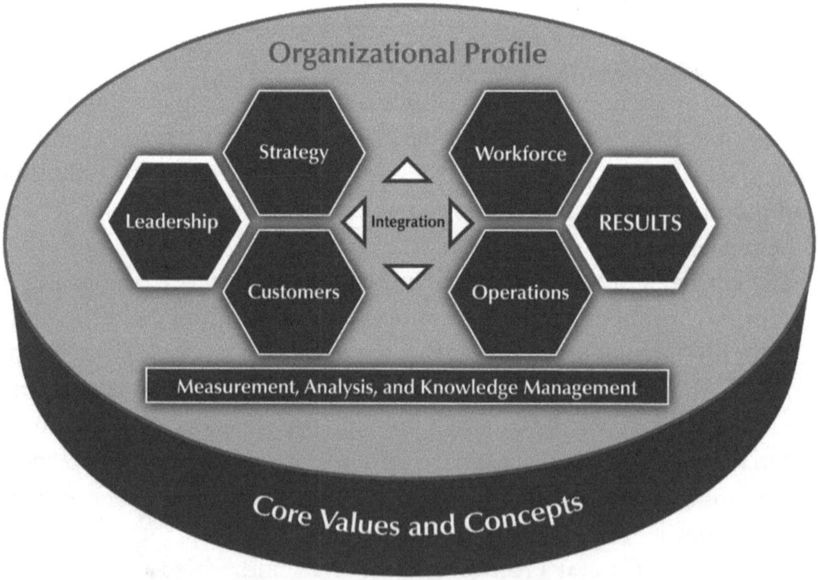

From Baldrige Performance Excellence Program. 2015. *2015–2016 Baldrige Excellence Framework: A Systems Approach to Improving Your Organization's Performance*. Gaithersburg, MD: U.S. Department of Commerce, National Institute of Standards and Technology. http://www.nist.gov/baldrige.

Fig. 8.5 Illustration of the Baldrige Excellence Framework (also known as the 'Baldrige Puck')

The framework sets out *Core Values and Concepts* that represent the embedded beliefs and behaviours found in high-performing organisations (e.g. those set out earlier). These underpin the criteria, which can be defined as an 'integrated management framework'—a tool for understanding and managing organisational performance. They are a set of questions that guide how to run any organisation, no matter its sector or size. The core values and criteria are used to generate an *Organisational Profile.*

The criteria represent all the components of a performance management system and are used to assess an organisation's performance, helping the organisation identify its strengths, opportunities for improvement and gaps/blind spots.

The first three criteria are *Leadership, Strategy and Customers*. The premise is that leadership should be focused on strategy and customers and this sets the tone and vision for the organisation. *These criteria are referred to as the **leadership triad** and are considered by the Baldrige framework to be the key contributors to an organisations success.*

The fourth category is *Measurement, Analysis, and Knowledge Management*. These linked activities provide a "fact base" that allows for data and knowledge-driven decision-making in the organisation. *This aspect of performance management enables all the other functions in the organisation to deliver successful outcomes.*

The final three criteria are *workforce, operations and results*. The results the organisation achieves are the outcome of the people (workforce) and processes (operations) that produce the organisation's goods and services, implement the organisation's strategy and serve its customers. *These criteria are called the **results triad.***

The Baldrige framework emphasises that the performance management system must be focused on leadership *and* results for the organisation to have ongoing success. In the scoring system, leadership and results have more points assigned than any of the other categories. It also emphasises that integration is at the core of the framework—the parts of an effective performance management system are interdependent and the performance of whole system is greater than the sum of individual parts. Organisations can evaluate their performance and level of maturity against the Baldrige Criteria (Fig. 8.6).

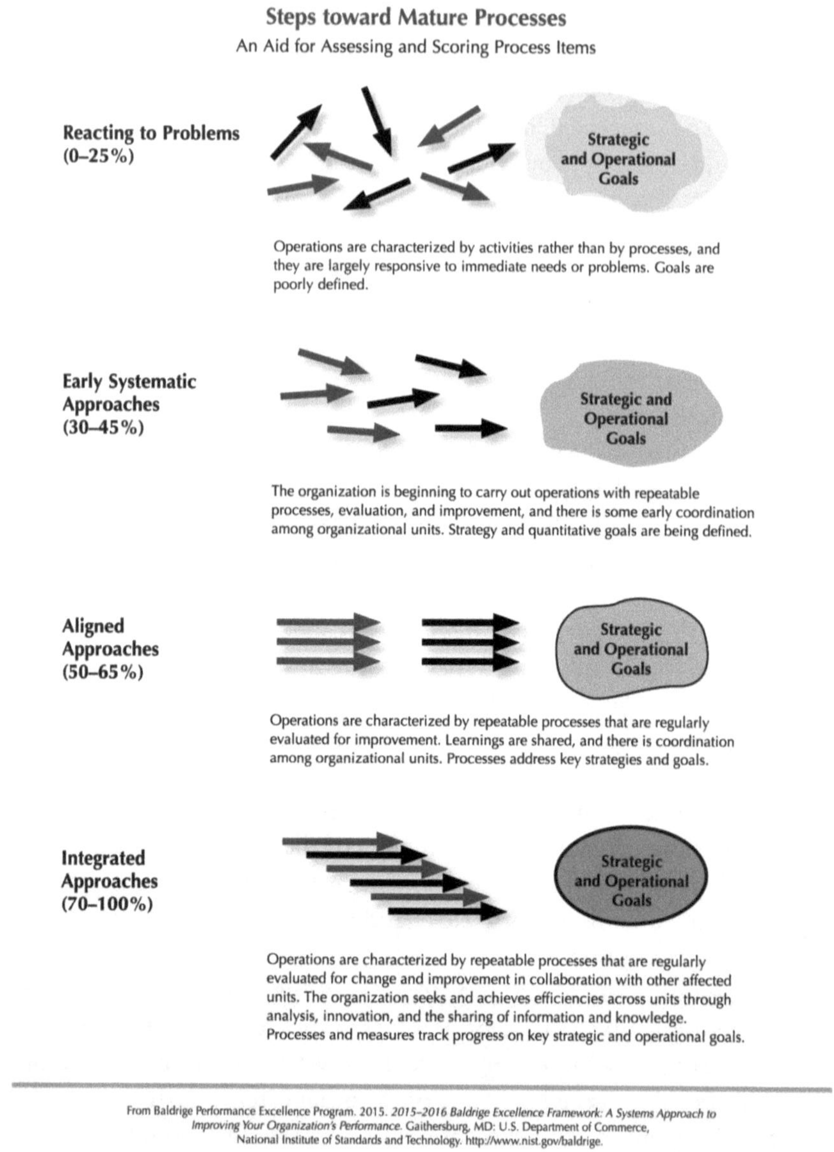

Fig. 8.6 The level of maturity of an organisation against Baldrige Criteria

Chartered Quality Institute Management System Specification and Guidance MSS 1000:2014

MSS 1000:2014 is a universal management system standard (MSS) enabling organisations to create fully integrated management systems directing and guiding their strategic, tactical and operational management processes. The MSS has been created by a team of experts within the CQI Integrated Management Special Interest Group (CQI, referenced 2016).

The MSS was written to:

- Demonstrate that a universal MSS is possible and practicable
- Act as a robust foundation to stimulate further research, innovation and continual improvement in the practice of integrated management
- Deliver the organisation's purpose through integrated structures and processes rather than attempting to independently manage multiple dimensions of performance

Implementing MSS 1000 is intended to significantly improve:

- Stakeholder satisfaction while making the best use of resources
- The effectiveness, efficiency and control of the organisation's structures and processes
- Prospect and risk management
- Commercially responsible and socially responsible performance
- The organisation's robustness, agility and resilience

Figure 8.7 illustrates the elements an organisation needs to manage across its all its internal and external stakeholder requirements (its context). For environmental management to be integrated into overall management in a meaningful and transparent way, the focus of management should be on those stakeholders, structures and processes which have a significant impact on the environment.

The MSS 1000 specification is based on universal management principles and definitions to enable organisations to manage in an integrated way. This has the effect of shifting the focus of management from multiple discrete facets of performance, such as goods and services quality,

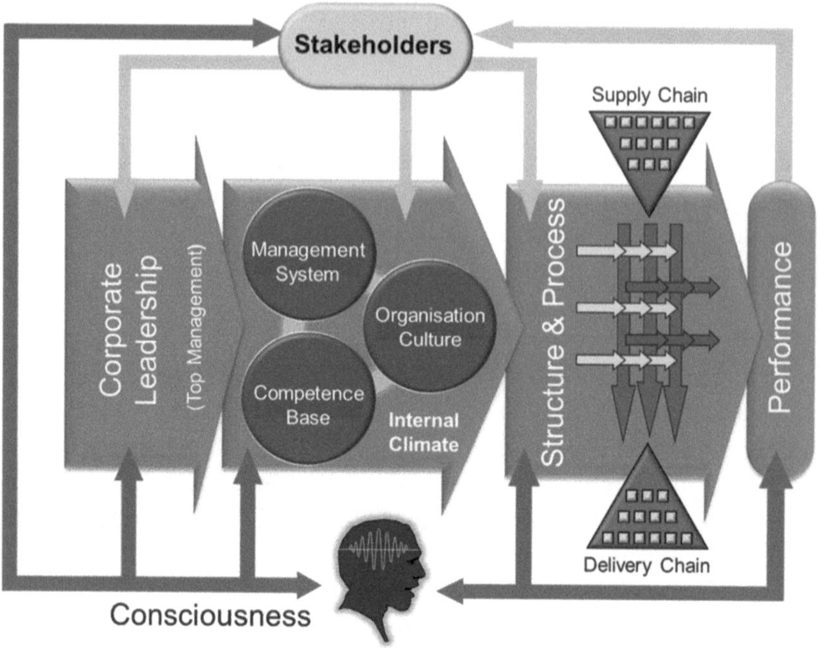

Fig. 8.7 Context of a management system (MSS 1000:2014)

people protection and nurture, environmental protection, security and so on, directly onto the organisation's structures and processes, which is where all aspects of performance emerge. Concepts such as aspect and impacts have been defined universally and not restricted to just environmental management (as is the case with ISO 14001, the standard for EMSs). This is illustrated in Fig. 8.8.

8.4 Conclusions

1. Development of environmental regulation

- Environmental regulation must adapt and evolve by moving from approaches based on instruction and sanctions to approaches that reflect the ways that successful organisations operate and interact in their 'working' environment.

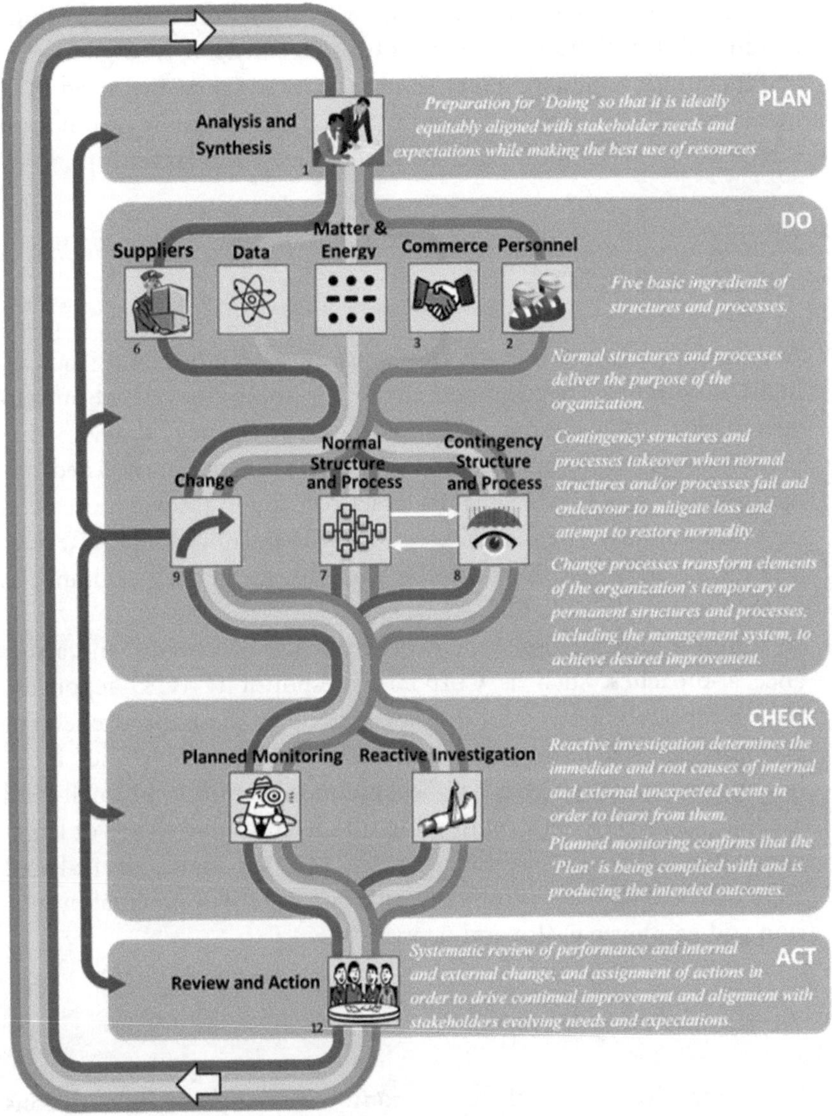

Fig. 8.8 Universal plan-do-check-act 12-element structure (Fig. 4 in MSS 1000:2014)

- Environmental management cannot be meaningfully and successfully conducted in isolation of the overall management of an organisation and its multidisciplinary interactions. The interactions that need to be managed include direct interactions on the environment and indirect interactions via economic and social activities and relationships.

2. Securing compliance (as a starting point for the 'journey' to effective risk management)

- As government (state) resources are constrained, it is becoming increasingly important to develop and encourage approaches based on self-governance and assurance. The Environment Agency (England) has been encouraged to increase the use of self-assurance and earned recognition schemes in its regulatory practices, and to recover more of its costs from industry, by the Cabinet Office's Regulatory Futures Review. The review adds that there may be scope for greater self-assurance as EMSs improve (ENDS 2017).
- In addition to regulation benefitting from self-governance and assurance approaches, such as Corporate Responsibity (CR) reporting, regulation can be a driver for increased uptake of these approaches. For example, as CR reporting rates increase in the largest companies, it can be expected that reporting will become more common in small- and medium-sized companies within the supply and value chains of larger companies. This has already been the case for management-based approaches to improve product/service quality, environmental protection and employee health and safety.

3. The role and use of interventions

- In an imperfect world, there is a continuing need for interventions from the state, and within value chains, to correct market failures and to influence/drive better behaviour and practice (as described above). There may need to be more emphasis on interventions that 'nudge' than those that 'prescribe' to effectively change the behaviour of organisations and individuals.

- The choice and effective use of interventions can be assessed using modelling techniques such as logic and dependency modelling.

4. The role and use of management systems, standards and specifications

- The development and use of management systems, specifications and standards can help organisations to make the transition from reacting to imposed regulation to proactive management.
- A fully integrated management system should cover all aspects of performance such as producing goods and services to specified quality, managing risks and opportunities, and managing stakeholder engagement. This is useful for resolving competing priorities, objectives and resources (e.g. quality of products/services, the requirements of investors, wishes of shareholders).

5. Developing and using innovative and integrated approaches

- Natural capital and ecosystem services, integrated management systems/specifications, and business performance frameworks provide approaches for an organisation to manage interactions, risks and opportunities within their value chains.
- Every organisation will have a unique set of interactions which it needs to manage, and it should decide what approaches and techniques work best for its own set of circumstances.

Notes

1. 319,324 certificates for ISO 14001 were issued worldwide in 2015, compared to 1,03,3936 certificates for ISO 9001, the standard for quality management systems (ISO Survey 2016).
2. Environment Agency staff numbers have been reduced from 12,500 in 2008 (England and Wales) to 9600 in 2016 (England) (Environment Agency 2008/9 and 2015/16). This includes approximately 1000 staff transferred to Natural Resources Wales in 2013 (Environment Agency Wales 2012).

References

Baldrige Performance Excellence Program. 2015. 2015–2016. *Baldrige Excellence Framework: A Systems Approach to Improving Your Organization's Performance (Business/Nonprofit)*. Gaithersburg, MD: U.S. Department of Commerce, National Institute of Standards and Technology. https://www.nist.gov/baldrige. See Also: Description of framework using text from web-site and graphics. http://www.baldrige.org/foundation/Mission-History.aspx

Chartered Quality Institute (CQI). 2016. *Management System Specification and Guidance MSS 1000:2014*. http://www.thecqi.org/Community/Special-Interest-Groups-SIGs/Integrated-Management-Group/Research-and-reports/. Extracted text to summarise MSS 1000: 2014 Specification and Guidance, including Figure 2 in MSS 1000: 2014 and Figure 4 in MSS 1000: 2014.

CQI Special Interest Group (SIG). 2016. *Integrated Management Systems Referenced from Web-Site*. http://www.thecqi.org/Community/Special-Interest-Groups-SIGs/Integrated-Management-Group/Research-and-reports

Defra. 2013. *Instrument Selection Guidance*. http://randd.defra.gov.uk/Document.aspx?Document=11687_InstrumentSelectionGuidance041113-external.pdf

ENDS. 2017. *Report on Cabinet Office's Regulatory Futures Review*. http://www.endsreport.com/article/55090/environment-agency-encouraged-to-increase-self-assurance

ENDS Report. 2016. *Change is Inherent in Corporate Reporting*, July 25. http://www.endsreport.com/article/53324/change-is-inherent-in-corporate-reporting. Selected text from article based on interview with Paul Scott, managing director of Corporate Register.

Environment Agency. 2008. *Effective Inspection and Enforcement: Implementing the Hampton Vision in the Environment Agency*. https://www.nao.org.uk/wp-content/uploads/2008/03/EA_Hampton_report.pdf. Text in quotes.

———. 2014. *Operational Risk Appraisal Methodology*. https://www.gov.uk/government/collections/operational-risk-appraisal-opra. Description of Opra attributes is author's personal interpretation of online content.

———. 2016a. *Operator Self-monitoring Scheme: MCERTs*. Selected text from web-page https://www.gov.uk/government/collections/monitoring-emissions-to-air-land-and-water-mcerts

———. 2016b. *Pig and Poultry Assurance Scheme*, Selected text from web-page https://www.gov.uk/government/publications/pig-and-poultry-assurance-scheme-intensive-farming

Environment Agency Annual Reports and Accounts 2008/9 and 2015/16. https://www.gov.uk/government/collections/environment-agency-annual-reports-and-accounts. Quoted figures from the report rounded by the author to nearest 100.
European Environment Agency. 2016. http://www.eea.europa.eu/
Hampton Review of Progress of Environment Agency, 2010. Selected quotes from the report.
HMT. 2016. *The Magenta Book*. https://www.gov.uk/government/publications/the-magenta-book
Environment Agency Wales. *Corporate Plan 2010–15*. http://abetterenvironmentforwales.co.uk/6004_EA_Eng%20index%20page_2.swf
IEMA. 2016. Environmentalist Journal, *Emisoft Statoil Case Study*. Emisoft and Statoil Case Study in Environmentalist Selected text agreed with Emisoft Project Manager, Inki R Brown. http://view.m.lexisnexis.co.uk/?qs=b786d736a397e4978ef2914322c881fc7ab939b0b4952b2c0129efb1e437d48c70ad3a25cf344fb24d7e11be4c090fb89910c722e095aa91102d4a7eed393affd596e206661774b0
IMPEL. 2006–2009. *Doing the Right Things Methodology*. http://www.impel.eu/projects/doing-the-right-things-methodology
———. 2014. *Supporting the Implementation of Integrated Risk Assessment*. http://www.impel.eu/projects/supporting-the-implementation-of-the-integrated-risk-assessment-method-iram
———. 2016. *Consultation Response to European Co-operation for Accreditation Regarding the Review of Guidance Document, EA-7/04—Legal Compliance as a Part of Accredited ISO 14001: 2015 Certification, 2016*. Text in Quotes from Personal Communication with IMPEL Project Manager.
IMPEL Report: 2014/12 April 1, 2015. *Choosing Appropriate Interventions, Phase 3*. Selection of Text from Project Report and Project Managers (the author) Personal Notes. http://www.impel.eu/projects/choosing-appropriate-interventions-phase-3.
IMPEL Report: 2015/18, November 25, 2015. *Implementation of the iDepend Decision Support Tool*. http://www.impel.eu/projects/implementation-of-the-idepend-decision-support-tool. Selection of Text from Project Report, User Guidance Document and Project Manager's (the author) Personal Notes. http://www.impel.eu/projects/implementation-of-the-idepend-decision-support-tool.
IMPEL Report: 2015/18, November 25, 2015. *Implementation of the iDepend Decision Support Tool*.
IMPEL Reports: 2011/04—2013/15—2014/16—2015/19—2016/19. *Compliance Assurance Through Company Compliance Management Systems*. Selection of text

from project reports and author's personal notes. http://www.impel.eu/projects/compliance-assurance-through-company-compliance-management-systems. See also http://www.impel.eu/tools/guidance-cms-assistance; http://www.impel.eu/projects/working-together-public-and-private-supervisor-assessing-ems-cms/.

ISO Survey. 2016. *The ISO Survey of Management System Standard Certifications 2015*. Statistics from Table Showing 2015 Survey Results. http://www.iso.org/iso/the_iso_survey_of_management_system_standard_certifications_2015.pdf

KPMG. 2015. *Currents of Change Survey of Corporate Responsibility Reporting 2015*. https://home.kpmg.com/xx/en/home/insights/2015/11/kpmg-international-survey-of-corporate-responsibility-reporting-2015.html

Natural Capital Coalition. 2016. Description of Protocol Including Graphic at Figure 5. http://naturalcapitalcoalition.org/protocol/

SNIFFER Report ER34. 2013. *Compliance Enforcement Model, Environment Agency*. Graphic Reproduced from Figure 1, Page 52 of Report https://www.sniffer.org.uk/files/5013/6749/5161/ER34_Project_Report_FNL.pdf

Viridor. 2011. *Viridor Openspace Web Portal Developed with Environment Agency*. Selected Text Used from Viridor/Environment Agency/EHS Data Brochures. http://ehsdata.com/documents/uploads/Viridor_CS.pdf. see also http://www.monitor-pro.co.uk/our-products/environment-agency-portal.aspx

Duncan Giddens worked for over 25 years as an environment and business professional in central government, the Environment Agency and the European Implementation and Enforcement of Environmental Law (IMPEL) network. Duncan has worked extensively on developing approaches to bring together environmental protection and sustainable business practice. He is now an independent consultant. Duncan's skills and expertise include the design and application of environmental and integrated management systems, environmental standards, risk appraisal and dependency modelling. He is a member of three UK National BSI Committees covering environmental management, environmental auditing and environmental performance evaluation. Duncan's additional areas of expertise and interest include corporate reporting, certification and accreditation to international standards and use of business excellence models.

9

The Impact of Leaving the European Union and the Future of Environmental Regulation

Martin Bigg

9.1 Introduction

The UK's membership of the European Union (EU) and its predecessors has had a major impact on the environmental regulation of UK businesses, as well as its ability to trade, and the sharing of information on environmental practice and performance across the EU. The UK has also significantly influenced the development and implementations of EU environmental regulations. There is a high level of consistency in regulation across Europe and this, as well as confidence and transparency in the regulations, has benefitted business, society and the economy. There are also areas of EU environmental policy and regulation which are not in the best interests of the environment. The separation of the UK from the rest of Europe gives opportunities for making improvements as well as posing threats to existing and future regulation.

M. Bigg (✉)
University of the West of England (UWE), Bristol, UK

J. Foreman (ed.), *Developments in Environmental Regulation*, Palgrave Studies in Environmental Policy and Regulation, DOI 10.1007/978-3-319-61937-8_9

9.2 Uncertainty

The one certain consequence of the UK referendum vote on 23 June 2016, 51.8% leave to 48.2% remain, was uncertainty. We did not know what our future relationship with the EU would look like nor how long, if ever, it would take to make the transition to a new relationship and how events would unfold during the transition.

Before the referendum there was a high level of understanding in the UK of the existing and developing environmental regulatory framework, which resulted from UK and EU legislation. Sustainability and commitment to the environment are at the heart of the EU. Article 3.3 of the Treaty on European Union and the Treaty on the Functioning of the European Union states:

> The Union shall establish an internal market. It shall work for the sustainable development of Europe based on balanced economic growth and price stability, a highly competitive social market economy, aiming at full employment and social progress, and a high level of protection and improvement of the quality of the environment. It shall promote scientific and technological advance.

After the referendum the future of existing and pending environmental legislation has been thrown into doubt. Even before any legislative changes, there is increasing uncertainty around which legislation will be applied in the future. This is driven by the divergence of views on what leaving the EU could mean for environmental regulation. These include scrapping renewable energy targets (Lilley 2016) or removing financial support for farming (House of Commons Library 2016) (Box 9.1).

Box 9.1 Common Agricultural Policy (CAP)

It is widely recognised that the CAP has not delivered for the environment and is overdue for review, particularly the protection of soils and water catchment (Parminter 2016). However, any resources released need to go to make farming more sustainable rather than replenishing the depleted UK exchequer. CAP funding of just under £3billion a year accounts for approximately

60% of farming income. Without EU farming controls, and access to EU markets, UK farming could become more intensive and polluting as the industry strives to survive and replace lost income.

Joan Walley Chair of the Aldersgate Group stated that she believed Brexit provided a "huge opportunity" to reform public payments to agriculture in a way that is supportive of both enhancing the environment and sustaining farm incomes. "Brexit will see the creation of a new support system in the UK and it should include increased incentives for sustainable land use practices that deliver environmental enhancements, such as improved water quality, flood protection and reductions in greenhouse gas emissions" (Ends 2017a).

9.3 Possibilities

The benefits of the collaborative development of EU and UK environmental policies and regulations have been widely recognised (House of Commons Environmental Audit Committee 2016). The EU has implemented a wide range of environmental directives and regulations as well as entering into many treaties and agreements on behalf of its members which are reflected in UK legislation (Odermatt 2016; Blockmans and Van der Loo 2016). The future of the UK's environmental regulations and their implementation will be determined in significant part by the nature of the UK's future relationship with the EU.

A further consideration is the relationship and regulatory responsibilities of the UK Westminster government compared with the devolved administrations. The Scottish government is committed to continue to uphold and contribute to EU environmental policy and legislation (Scottish Government 2016). The Northern Ireland and Irish governments have common environmental issues and the Welsh government is developing a differing environmental agenda to England (National Assembly for Wales 2016).

The UK government triggered the start of divorce proceedings from the EU by invoking Article 50 of the EU Lisbon Treaty states on 29 March 2017.

Article 50 states:

> Treaties shall cease to apply … from the date of entry into force of the withdrawal agreement or, failing that, two years after the notification… unless the European Council, in agreement with the Member State concerned, unanimously decides to extend this period.

On this timetable, the UK is due to be out of the EU by the end of March 2019.

Initial options for the UK government included:

1. Membership of the European Economic Area outside the EU, with Iceland, Liechtenstein and Norway
    a. Access to the EU single market
    b. Compliance with the four freedoms (freedom of movement of goods, services, money and people)
    c. Most EU law applies
    d. Some directives will not apply, for example, the Birds and Habitats Directive
2. Membership of the European Free Trade Area outside the EU, with Iceland, Liechtenstein, Norway and Switzerland
    a. Access to a worldwide network of free trade and partnership agreements
    b. Compliance with EU laws by replicating in national legislation
3. Bespoke trade deal
    a. EU law would apply to imports to single market
    b. Other EU laws would not apply
4. Membership of the World Trade Organisation (WTO) only
    a. The UK has been a member since 1 January 1995

On 17 January 2017 at Lancaster House in London, the UK Prime Minister Theresa May announced the government's plan for negotiating Brexit including 12 priorities. The details were set out in a white paper published the following month (HM Government 2017). The UK government confirmed that it intended to fully withdraw from any form of membership of, or association with, the EU. It wanted:

1. Access to the single market through a new, comprehensive, bold and ambitious free-trade agreement
2. Customs arrangements which avoided delays and paperwork
3. Full control of immigration but an openness to targeted European migrants in specific sectors, with the rights guaranteed of long-term British residents in Europe and EU residents in Britain
4. Continued participation in European security and research organisations
5. No more "vast" annual payments into the EU budget
6. A phased movement into the new arrangements, with different periods of transition in different sectors

The government also stated that it wanted a free hand to negotiate future arrangements with the EU without parliamentary scrutiny.

It took the EU and Canada, with a high level of commitment, seven years to agree a deal on trade and investment even before it was eventually ratified by all EU member states (after a few last-minute delays). Recognising the starting postures of UK and EU governments, a trade deal between them could take longer. With the new US President in 2017 apparently committed to tearing up trade deals, the future of international agreements looked poor. If necessary, the UK will leave the EU without any agreements in place and rely on World Trade Organisation rules in the conduct of its international business.

There is uncertainty as to what if any agreements can be reached or whether they can be completed in the two-year time window. There is also no certainty as to whether the time window can be adhered to or extensions sought or given.

The UK government has indicated that it wants to resist the freedom of movement of people while retaining the other freedoms and to "take back control". However, the leaders of the other member states in the EU and the European Parliament have consistently restated that participation in the single market requires the adoption of the four freedoms as described in the EU rules (HM Government 2016; Financial Times 2016a).

Whichever of the outcomes, if the UK is to continue to trade with the remaining members of the EU (accounting for 47% of UK exports in August 2016), it is likely that the UK will be required to comply with some or all of the EU environmental legislation and contribute to EU funding. This will be without having any significant influence in its operation or future development (HM Revenue and Customs 2016). With or without the UK, the EU will continue to develop and revise environmental regulations, so in order to continue to trade with the EU the UK will also have to continue to implement parallel legislation.

9.4 Regulations

Most EU environmental laws apply to not only the 28 EU member states but also to the additional members of the European Economic Area (Iceland, Liechtenstein and Norway). Until the completion of the implementation of Article 50, the UK remains a full member, has to comply with all the environmental laws and should be fully involved in all discussions and negotiations. However, there was no guarantee that this will happen and, as of 29 March 2017, it appeared that UK nationals were being excluded from taking leading roles in the development of any new legislation.

Within weeks of the referendum, concerns were raised regarding the status of EU nationals working in the UK as well as UK nationals working in the other EU countries. This has impacted on the development of collaborative projects and future research, with UK participation becoming less welcome. It is now seen as less important to have a UK contribution to the development of future European research, policies, laws, standards and guidance. Even where the UK government and UK representatives participate in European negotiations, there is a tendency to pay less attention to them.

Important regulations which required implementation in the interim period included the revised National Emissions Ceiling Directive which had to be transposed by 1 July 2018. It sets emission reduction commitments in relation to certain atmospheric pollutants for member states and requires them to draw up, adopt and implement national air pollution control programmes as well as to monitor and report on emissions levels of the pollutants in the directive.

The Medium Combustion Plant (MCP) Directive for facilities between 1 and 50 MW sets emissions limit values for pollutants including sulphur dioxide, nitrogen oxides and dust, as well as monitoring and assessment requirements. The MCP has to be transposed into UK law before December 2017. Because of the increase in small power plant in the UK, especially diesel generators, used at short notice, there is a strong incentive not only to meet the standards in the directive but also to go beyond them. This will be a test of the UK government commitment to improving air quality rather than reducing regulation (Box 9.2).

Box 9.2 Anticipating leaving the EU?

EU air-quality regulation has helped raise awareness of the estimated 40,000 early deaths in the UK every year, and the cost to the economy of £24bn, due to bad air quality. The UK has been in breach of EU air pollution limits for the pollutant nitrogen dioxide (NO_2) in 16 of 43 zones—limits which the UK was required to meet in 2010 under the EU's Ambient Air Quality Directive (European Commission 2017a).

The European Commission initiated infringement action and the UK Supreme Court has ruled twice that the government's actions were inadequate (New Scientist 2016). As its previous air quality plan was too weak and timed to avoid infringement proceedings, the government was ordered to produce a new plan detailing actions to reduce NO_2 levels and comply with EU limits. It has produced a slim plan which placed the principle responsibility on local authorities and is seen by many as woefully inadequate. A sceptic might wonder, based on past performance, if the government's approach is designed to delay action until the UK had left the EU (National Audit Office 2014).

The UK government has also pushed the European Commission to delay tightening pollution limits emissions from large combustion plants as required under the Industrial Emissions Directive (Financial Times 2016b).

Without the jurisdiction of the European Court of Justice as a last resort, there is also a danger that UK standards may not be enforced effectively.

However, the Air Quality Directive is likely to be updated with new air quality limits in 2018. The UK limits could end up lagging behind. The EU could make continued compliance with air quality rules a condition of any UK–EU trade deal. Air pollution does not recognise international boundaries.

Across the corridors of Brussels over the past 44 years, the common language had shifted from French to English and the culture and ways of working have become more English/western European. It surprised the UK government and UK media that the negotiations could at least in part be conducted in French (Reuters 2016).

The UK government has proposed a "Great Repeal Bill" to repeal the European Communities Act 1972 from the moment the UK leaves the EU (Douglas-Scott 2016; Conservative Party 2016). It will also put the current EU legislation into UK law. This is a serious challenge to a significantly depleted UK civil service.

It was reported in *The Times* newspaper on 24 October 2016 that the then Minister for International Development Grant Shapps was planning to attach a "sunset clause" under which the EU-made laws within the bill would stop being in force after five years. Mr. Shapps was concerned that the bill would in reality be the "Great Continuity Bill" because it would transfer all of the EU legislation into the British law. *The Independent* newspaper on 25 March 2017 reported that Michael Gove, a key figure in the Brexit campaign, said the pledge to "take back control" could allow Britain to scrap "absurd" rules such as the European Commission's Habitats Directive and Clinical Trials Directive (Independent 2017).

The then Environment Secretary, Andrea Leadsom, giving evidence to the House of Commons Environmental Audit Committee (EAC) inquiry into the impact of Brexit on 25 October 2016, said that around one-third of EU environmental legislation will be difficult to transfer directly into UK law. New laws or mechanisms would be needed to implement the various items of legislation. Around 80% of UK environmental legislation was estimated to originate from the EU, including regulations covering waste and recycling, air and water quality, chemicals, biodiversity, energy efficiency, climate change, fisheries and agriculture, and clean energy.

When the white paper on the bill to repeal the European Communities Act 1972 was published on 30 March 2017, it was clear that the bigger task was to copy onto the statute book the vast and complex web of European regulations (Department for Exiting the European Union 2017). The paper reiterated an earlier government commitment to "ensuring that we become the first generation to leave the environment in a better state than we found it". Claiming that UK's legislative framework has delivered tangible environmental benefits "such as cleaner rivers and reductions in emissions of sulphur dioxide and ozone-depleting substances emissions", it stated that the Great Repeal Bill will ensure that the whole body of existing EU environmental law continued to have effect in UK law. This was in marked contrast to the aspirations of some members of parliament as mentioned above.

The white paper stated that "a very significant proportion of EU-derived law for which UK government departments are responsible contains provisions that will not function appropriately if simply copied over". About 800–1000 pieces of legislation needed to be adjusted within the next two years so that they make sense in a post-EU UK. To achieve this and without getting tied up in parliamentary process, ministers proposed to use secondary legislation. Ironically, it was this reduced parliamentary scrutiny or control that drove some members of parliament to support leaving the EU.

An analysis by Ends has identified several ways in which the legal relationship between UK and the EU could begin to loosen even before the final date of separation (Alice Fillan 2017). In addition to EU directives which set objectives that the UK converted into its own laws, there are the regulations which were directly applicable. They apply immediately to the UK without any national legislation. Ends reported that there were about 500 directly applicable regulations with environmental effects in force.

EU and UK legislation refers to EU wide guidance, definitions, standards and interpretations. This cross-referencing also needs to be covered in the new legislation. Much of this material is the result of extensive international and national debate and agreement but is very vulnerable to individual challenge. Dispensing with the "red tape" was a strong selling point of the campaign to leave the EU, so it is unlikely that this material supporting EU legislation will survive unscathed.

At the same time, UK legislation needs to be kept in tune with developing EU legislation in order to maintain the chosen new UK relationship with the EU.

Transposition is further complicated by the different legislative frameworks in the devolved administrations. One possibility could have been to concentrate the transitions in Westminster and devolve them subsequently. This would have not gone well with the devolved administrations and their different agendas towards EU engagement. Any legislative changes will also have taken time to embed, typically 18 months, as regulators and regulated interpret and test the wording of new legislation.

A further challenge is the enforcement of environmental regulation. The European Court of Justice will no longer be able to hold the UK government to account or bind UK courts. Only the Supreme Court and the judicial review process will be able to hold the UK government and public bodies to account. UK judges will continue to be bound by European case law up to the point of leaving, but new judgments of the Supreme Court will be able to supersede European rulings.

The UK withdrawal from Euratom is likely to result in the UK creating a new UK nuclear regulator (Prime Minister's Office 2017). As the UK government is committed to breaking from the jurisdiction of the European Court of Justice, it has to develop a new way of regulating chemicals (Box 9.3).

Box 9.3 EU registration, evaluation, authorisation and restriction of chemicals (REACH)

The UK chemical and related industries are strong supporters of the EU REACH regulations and their continuing enforcement backed by a Europe-wide body (Environmental Audit Committee 2017). For the UK industry to be able to trade with the EU, it will have to follow the REACH registration, authorisation and restriction process. However, the UK cannot continue to be in REACH because it is overseen by the European Court of Justice. The UK will therefore have to have its own working chemicals regime from the moment the UK leaves. The cost to the UK of the additional work could be tens of millions of pounds.

9.5 Opportunities

Fortunately, despite the uncertainty and confusion following the vote to exit the EU, across most businesses there is recognition of the need for consistent, effective and enforced environmental regulation. The UK still needs to comply with international treaties and conventions. Environmental issues cross national boundaries so international collaboration continues to be required. Approaches to environmental regulation have to:

- Be precautionary
- Be technically feasible
- Take account of the costs and benefits to the business sector and wider society
- Focus on outcomes
- Take account of the total environment: air, land and water

International collaboration will still be required to develop standards, assess techniques and agree enforcement measures.

There could be an opportunity to consolidate environmental regulation, apply a more consistent approach to pollution of air, water and land, and the management of waste, to secure the best solution for the environment as a whole. Unnecessary bureaucracy, measuring, inspection and reporting continues to need challenging, with more focus on exception reporting and auditing, targeting the persistent non-compliant and deviant. These aspects could be included in the government's 25-year environment plan, publication of which appears to be perpetually delayed.

Support for energy needs to be more sustainable with greater emphasis on environmental protection, renewable energy and long-term planning and investment. Past EU support for low carbon and environmental innovation needs to be enhanced, not reduced.

However, the UK government believes that a new comprehensive free-trade deal is possible, precisely because the UK and EU regulations were equivalent. While that could be the case at the point of departure, if UK politicians change regulations either as part of a deregulation or new regulation initiative, it will not be long before there will be significant divergence and free-trade agreements will be off the agenda.

9.6 Future Regulation Strategy

In January 2017, the UK government sought views on its approach towards building a modern industrial strategy (Department for Business, Energy and Industry Strategy 2017). The overall objective of the strategy was "to improve living standards and economic growth by increasing productivity and driving growth across the whole country". It contained a chapter on delivering affordable energy and clean growth but was weak on delivering sustainable development, resource efficiency or improving environmental performance.

UK energy prices are directly linked to the European energy market and the trade through the various interconnectors. New energy agreements will be required with the EU as well as the application of World Trade Organisation rules increasing prices. Increased uncertainty may result in increased demand to retain old energy generating plants and increase the use of less efficient fossil fuel intensive engines. In the future, the public and environmental regulators will need to be watchful that security of supply and price did not trump environmental standards and the achievement of climate change commitments.

The European Parliament has drawn a line in the sand on the environment and climate change which could provide a possible level of protection for the future (European Parliament 2017):

> …any future agreement between the European Union and the United Kingdom is conditional on the United Kingdom's continued adherence to the standards provided by international obligations, including human rights, and the Union's legislation and policies, in, among others, the fields of the environment, climate change, the fight against tax evasion and avoidance, fair competition, trade and social rights, especially safeguards against social dumping.

An indication of the future direction of regulation in England was given in March 2017 when 13 sector strategies for industries regulated by the Environment Agency were published (Environment Agency 2017). They set out the actions to be undertaken between 2016 and 2020. The strategy documents were dated 2016 and some actions had already been completed. The strategies focused on inspection and enforcement. Tackling climate change varied with more references in the waste sector. Several strategies referred to the delivery of emission reductions through the timely implementation of the "Best Available Techniques Reference (BREF)" conclusions, and compliance with EU legislation including the Industrial Emissions Directive and the Water Framework Directive, until the UK leaves the EU.

The delay in publication, continuation of existing actions, tackling continuing non-compliance especially in the waste sector, depending on EU directives and standards, reflected the loss of resources and support in the Environment Agency. The lack of work on the development of guidance and standards, and providing advice to business, will be expected to have implications for future environmental protection. This is well demonstrated in waste management. The dramatic reduction in local authority funding has resulted in the closure and reduced working of civic amenity sites. Fly-tipping and illegal waste activities have increased, resulting in additional costs to industry and local authorities (Department for Environment, Food and Rural Affairs (Defra) 2017a). Waste recovery and recycling is not improving as much as it used to, hitting the businesses which provide the service. Meanwhile the European Parliament and Commission are tightening up recycling targets (European Commission 2017b).

The retrenching of environmental regulation in England is in stark contrast with the more proactive and engaging approach to regulation in Scotland and Northern Ireland. These two countries provide detailed sector-specific environmental guidance for businesses through their NetRegs web portal (Scottish Environment Protection Agency, Northern Ireland Environment Agency 2017). Everything from legislation, air pollution, emergency responses to nuisance is included in a readily accessible form. Ironically, NetRegs was set up in partnership with the Environment Agency in England but was an early victim of the centralisation of policy and guidance in Defra. The Welsh government has worked hard to improve environmental performance by businesses. It had ambitious plans to move towards a circular economy and become a "zero-waste nation" by 2050. It was a partner in EU-funded circular economy projects including one for SMEs (Interreg Europe 2016).

With increasing devolution, it is very likely that the different countries will take increasingly different approaches to environmental regulation.

A positive indication for the future of regulation was the record fine in March 2017 on Thames Water of nearly £20m for repeatedly polluting the Thames and surrounding watercourses and land with raw sewage (Ends 2017d). This followed a series of non-compliances and prosecutions and should be compared with £2bn revenue in 2016. It will be interesting to see if such fines have an impact on businesses in the future. It will also be interesting to see if the regulators use their powers to take actions against individuals where there has been consent or connivance, a power currently used more with criminal activity in small waste businesses.

9.7 Future of Environmental Regulation

For a moment on 29 March 2017, the UK government was in control. It had set in motion the divorce from the EU. But from that moment, it started to lose control as the timetable and agenda moved into other hands. The EU has made clear that negotiations on the divorce would be separate to the negotiations on any future arrangements. Progress on changing UK legislation will be dependent on the different positions and interests of the members of the UK parliament. The consensus is that it could take significantly more than two years to sever links with the EU and put the necessary new UK legislation in place. It could take three or five years or longer—we do not really know.

During this interim period most legislation will remain in place. The investment in influencing EU legislation will become a sunk cost as businesses have to invest in new ways of working with and understanding the new legislation. The remaining 27 EU members will continue as major global market players, setting and shaping regulations and standards, while the UK becomes more of a policy taker rather than a policy-maker, beholden to EU, US or other trade requirements. The UK government will no longer have to comply with European rulings. The devolved administrations will increasingly take independent positions on contentious issues such as fracking, genetically modified organisms and air quality unless reined in by central government. Meanwhile UK government ministers may well secure major changes through statutory instruments, without parliamentary scrutiny, except where sharp-eyed experts catch out what is going on. Even then stopping the administrative machine will be very difficult.

In ten years, the UK will have started to adjust to the new world order, working with new trading arrangements dominated by the growing global players, its environmental controls determined by the needs of growth and trade. In 30 years, the major trading players are likely to be China, India, Russia, US and the EU. It will be interesting to see who the UK tries to join up with.

9.8 Conclusions

When the UK government triggered Article 50 on 29 March 2017, there was much uncertainty and concern. What was known was that the government's resources and staff were under great pressure and very stretched (Institute for Government 2017; Institute for Government 2016). At the EAC's inquiry into the government's climate change risk assessment, it found that three quarters of the staff in the team had left (Department for Environment, Food and Rural Affairs 2017b; Ends 2017b). This was part of a far bigger staff reduction across the environment department and agencies (Ends 2016, 2017c). This poses a potential precursor for mistakes and a limited ability to identify and tackle any other issues.

The Times newspaper in April 2017 reported the contents of a Department for International Trade briefing which reflected the pressures on resources and trade priorities: "Trade and growth are now priorities for all posts—you will all need to prioritise developing capability in this area. Some economic security-related work like climate change and illegal wildlife trade will be scaled down" (Times 2017).

On a positive note, the annual review by the Institute for Environmental Management and Assessment for 2017 reported that seven in ten Environment and Sustainability Professionals said they are satisfied or highly satisfied with their roles (Institute of Environmental Management and Assessment (IEMA) 2017). Further, eight in ten of those who have moved to the profession from another occupation were happy with their chosen career.

Action on significant issues is being deferred, such as addressing air pollution and the related issue of vehicle emissions. Policies, decisions and actions on the emissions reduction plan needed to deliver the fifth carbon budget are missing. It is unclear how the UK will interface with the EU ETS and its future development or respond to EU development of the circular economy and resource efficiency. Energy generators and users are hit by the uncertainty in the energy market, the future of coal-fired power stations and support for renewable energy (Box 9.4).

Box 9.4 Climate change and emissions trading

The Climate Change Act committed the UK to reducing its emissions by 80% by 2050 from 1990 levels. The UK remains party to the United Nations Framework Convention on Climate Change and signatory to the Paris Agreement. On leaving the EU, the UK has to review its commitment to the EU decarbonisation target, which also affected other EU states. However, the UK government cancelled its carbon capture and storage project, its low carbon strategy was delayed and agricultural emissions not addressed, putting the carbon budgets at risk.

The EU emissions trading scheme (ETS) is the biggest carbon market in the world and covers around 1400 installations across the UK—including factories, power plants, heavy industry sites—emitting around 145 MT each year, accounting for 8% of the scheme. It has generated significant funds for renewable energy and reduced carbon emissions across the EU. Leaving the ETS will impact not only these installations but also the EU-wide market under which they trade. While the EU ETS includes states outside the EU, it will be difficult for the UK to withdraw from the scheme due to the unsettling impact on industry and the UK carbon budgets.

Within the EU ETS there is a glut of emissions allowances and low carbon prices. International representatives have been working hard to address the problems before the next trading period starts in 2020. The UK has a choice as to whether to continue within the scheme until or post 2020, which includes the jurisdiction of the European Court of Justice, or risk further instability in the energy market. As the ETS is run by the European Commission, a new arrangement is likely to be required. In the transition period, it is assumed that the UK will continue to participate in existing arrangements (Businessgreen 2017).

When the government published its proposals on the future relationship of the UK with the EU, businesses, investors and the public sought clarity and confidence (HM Government 2017; Mayor of London 2017). However, the government was determined not to share anything for fear of jeopardising its negotiating position. The uncertainties are expected to continue well beyond the two years allowed for in Article 50.

The scale of the impact and benefits of Europe wide environment regulations is seen by many as an advantage of membership of the EU. The regulations are also seen as a constraint and burden by others. The separation of the UK from the rest of Europe will result in a loss of influence on the development of controls and standards, which will still apply to the UK, and many international treaties and obligations will continue to apply. The UK is challenged to retain the best of EU environmental regulations while striving to improve areas where there are deficiencies. To secure the future of environmental regulation the UK needs to ensure that it does not waste a good crisis.

References

Alice Fillan, M.L.N.R. 2017. *UK Environmental Law on the Way to Brexit.* Twickenham: Ends.

Blockmans, S., and G. Van der Loo. 2016. *The Impact of Brexit on the EU's International Agreements.* https://www.ceps.eu/publications/impact-brexit-eu%E2%80%99s-international-agreements. Accessed 7 Nov 2016.

Businessgreen. 2017. *Brexit and the EU ETS: Carving Up the Emissions Market.* http://www.businessgreen.com/bg/analysis/3007788/brexit-and-the-eu-ets-carving-up-the-emissions-market/page/2. Accessed 7 Apr 2017.

Conservative Party. 2016. *Prime Minister: The Good That Government Can Do.* http://press.conservatives.com/post/151378268295/prime-minister-the-good-that-government-can-do. Accessed 12 Nov 2016.

Department for Business, Energy and Industry Strategy. 2017. *Building Our Industry Strategy.* https://www.gov.uk/government/organisations/department-for-business-energy-and-industrial-strategy. Accessed 28 Mar 2017.

Department for Environment, Food and Rural Affairs. 2017a. *Fly-Tipping Statistics for England, 2015/16.* https://www.gov.uk/government/uploads/system/uploads/attachment_data/file/595773/Flytipping_201516_statistical_release.pdf. Accessed 28 Mar 2017.

———. 2017b. *UK Climate Change Risk Assessment 2017.* https://www.gov.uk/government/publications/uk-climate-change-risk-assessment-2017. Accessed 2017 Mar 2017.

Department for Exiting the European Union. 2017. *Legislating for the United Kingdom's Withdrawl from the European Union.* https://www.gov.uk/government/uploads/system/uploads/attachment_data/file/604514/Great_repeal_bill_white_paper_print.pdf. Accessed 4 Apr 2017.

Douglas-Scott, S. 2016. *The 'Great Repeal Bill': Constitutional Chaos and Constitutional Crisis?* UK Constitutional Law Association. Blog. http://wp.me/p1cVqo-1ev. Accessed 8 Nov 2016.

Ends. 2016. *DEFRA 'Unprepared for Brexit' After Slashing Staff in Referendum Run-Up.* http://www.endsreport.com/article/54789/defra-unprepared-for-brexit-after-slashing-staff-in-referendum-run-up. Accessed 29 Mar 2017.

———. 2017a. *Brexit 'Should Not Distract' from Other Government Priorities.* http://www.endsreport.com/article/55751/brexit-should-not-distract-from-other-government-priorities. Accessed 4 Apr 2017.

———. 2017b. *DEFRA Climate Adaptation Team Slashed by Three Quarters.* http://www.endsreport.com/article/55760/defra-climate-adaptation-team-slashed-by-three-quarters. Accessed 29 Mar 2017.

———. 2017c. *Environment Agency Slashes Frontline Enforcement Staff.* http://www.endsreport.com/article/56054/environment-agency-slashes-frontline-enforcement-staff?. Accessed 19 Apr 2017.

———. 2017d. *Thames Water Fined Record £20m for Repeated Raw Sewage Pollution.* http://www.endsreport.com/article/55353/thames-water-fined-record-20m-for-repeated-raw-sewage-pollution. Accessed 29 Mar 2017.

Environment Agency. 2017. *2015 Regulated Industry Sector Strategies.* https://ea.sharefile.com/share?#/view/sf4a50053b28440bb. Accessed 28 Mar 2017.

Environmental Audit Committee. 2017. *The Future of Chemicals Regulation After the EU Referendum Inquiry.* http://www.parliament.uk/business/committees/committees-a-z/commons-select/environmental-audit-committee/inquiries/parliament-2015/inquiry2/publications/. Accessed 29 Mar 2017.

European Commission. 2017a. *Commission Warns Germany, France, Spain, Italy and the United Kingdom of Continued Air Pollution Breaches.* http://europa.eu/rapid/press-release_IP-17-238_en.htm. Accessed 4 Apr 2017.

———. 2017b. *Report on the Implementation of the Circular Economy Action Plan.* http://ec.europa.eu/environment/circular-economy/implementation_report.pdf. Accessed 4 Apr 2017.

European Parliament. 2017. *Negotiations with the United Kingdom Following its Notification that It Intends to Withdraw from the European Union*. http://www.europarl.europa.eu/sides/getDoc.do?pubRef=-//EP//TEXT+TA+P8-TA-2017-0102+0+DOC+XML+V0//EN&language=EN. Accessed 19 Apr 2017.

Financial Times. 2016a. *Merkel Warns European Business Not to Press for 'Comfortable' UK Trade Deals*. https://www.ft.com/content/113754ac-8bc5-11e6-8cb7-e7ada1d123b1. Accessed 7 Nov 2016.

———. 2016b. *UK Calls for Delay to Stricter Coal Pollution Rules*. https://www.ft.com/content/31adfd52-96b6-11e6-a1dc-bdf38d484582. Accessed 8 Nov 2016.

HM Government. 2016. *Rights and Obligations of European Union Membership*. https://www.gov.uk/government/uploads/system/uploads/attachment_data/file/516501/Rights_and_obligations_of_European_Union_membership_web_version.pdf. Accessed 7 Nov 2016.

———. 2017. *The United Kingdom's Exit from and the New Partnership with the European Union*. https://www.gov.uk/government/uploads/system/uploads/attachment_data/file/589189/The_United_Kingdoms_exit_from_and_partnership_with_the_EU_Print.pdf. Accessed 29 Mar 2017.

HM Revenue and Customs. 2016. *Overseas Trade Statistics—Non-EU and EU Trade*. https://www.uktradeinfo.com/Statistics/OverseasTradeStatistics/Pages/EU_and_Non-EU_Data.aspx. Accessed 7 Nov 2016.

House of Commons Environmental Audit Committee. 2016. *EU and UK Environmental Policy, Third Report of Session 2015–16*. http://www.publications.parliament.uk/pa/cm201516/cmselect/cmenvaud/537/537.pdf. Accessed 7 Nov 2016.

House of Commons Library. 2016. *Briefing Paper Number 07213, Brexit: Impact Across Policy Areas*. http://researchbriefings.files.parliament.uk/documents/CBP-7213/CBP-7213.pdf. Accessed 7 Nov 2016.

Independent. 2017. *Slash EU Regulations on Wildlife Protection and Drug Safety Trials After Brexit, Michael Gove Urges*. http://www.independent.co.uk/news/uk/politics/brexit-eu-regulations-michael-gove-environment-drugs-a7649041.html. Accessed 19 Apr 2017.

Institute for Government. 2016. *Whitehall's Preparation for the UK's Exit from the EU*. https://www.instituteforgovernment.org.uk/sites/default/files/publications/IFGJ5327_Report_Brexit_Civil_Service_080317_WEB.pdf. Accessed 2017 Mar 2017.

———. 2017. *The Civil Service After Article 50.* https://www.instituteforgovernment.org.uk/sites/default/files/publications/IFGJ5327_Report_Brexit_Civil_Service_080317_WEB.pdf. Accessed 29 Mar 2017.

Institute of Environmental Management and Assessment (IEMA). 2017. *The State of the Profession, Practitioners Survey 2017.* https://www.iema.net/assets/newbuild/documents/State%20of%20the%20Profession%20survey.pdf. Accessed 19 Apr 2017.

Interreg Europe. 2016. *Circular Economy for SMEs.* https://www.interregeurope.eu/cesme/news/news-article/129/circular-economy-for-smes-project-kicks-off-in-wales/. Accessed 28 Mar 2017.

Lilley, P. 2016. *Daily Telegraph.* http://www.telegraph.co.uk/news/2016/08/31/we-should-seize-the-benefits-of-brexit-sooner-rather-than-later/. Accessed 7 Nov 2016.

Mayor of London. 2017. *London's Global and European Future.* https://www.london.gov.uk/sites/default/files/londons_global_and_european_future.pdf. Accessed 29 Mar 2017.

National Assembly for Wales. 2016. *Environment (Wales) Act 2016.* http://www.legislation.gov.uk/anaw/2016/3/pdfs/anaw_20160003_en.pdf. Accessed 7 Nov 2016.

National Audit Office. 2014. *Briefing for the House of Commons Environmental Audit Committee, Environmental Protection.* https://www.nao.org.uk/wp-content/uploads/2014/06/Environmental-Protection-briefing.pdf. Accessed 8 Nov 2016.

New Scientist. 2016. *Court Orders UK to Take Urgent Action to Reduce Air Pollution.* https://www.newscientist.com/article/2111117-court-orders-uk-to-take-urgent-action-to-reduce-air-pollution/. Accessed 8 Nov 2016.

Odermatt, J. 2016. *European Journal of International Law, Brexit and International Law.* http://www.ejiltalk.org/brexit-and-international-law/. Accessed 7 Nov 2016.

Parminter, B. 2016. *Brexit: What Does This Mean for the Environment?* London: Castle Debates.

Prime Minister's Office. 2017. *Prime Minister's Letter to Donald Tusk Triggering Article 50.* https://www.gov.uk/government/publications/prime-ministers-letter-to-donald-tusk-triggering-article-50. Accessed 4 Apr 2017.

Reuters. 2016. *Exclusive—Parlez-vous Brexit? EU Negotiator Wants Brits to Talk French.* http://uk.reuters.com/article/uk-britain-eu-language-exclusive-idUKKCN12L1EE. Accessed 8 Nov 2016.

Scottish Environment Protection Agency, Northern Ireland Environment Agency. 2017. *Netregs, Environmental Guidance for Your Business in Northern Ireland & Scotland.* http://www.netregs.org.uk/. Accessed 28 Mar 2017.
Scottish Government. 2016. *EU Environment Pledge.* http://news.gov.scot/news/eu-environment-pledge. Accessed 7 Nov 2016.
Times. 2017. *'Less Climate Concern' Key to Brexit Trade.* https://www.thetimes.co.uk/article/less-climate-concern-key-to-brexit-trade-g6lrs6cw3. Accessed 19 Apr 2017.

Martin Bigg is Professor of Environmental Technologies Innovation at the University of the West of England (UWE), Bristol, UK, and Chair of Council and Board Member of the Institute for Environmental Management and Assessment. He leads and supports environmental initiatives with governments, businesses, local authorities and communities. As the Director of the Environmental Technologies Innovation Network (iNet), he led a European programme to support innovative enterprising businesses. He was the Chair of the Bristol Green Capital Partnership, which secured the title of European Green Capital 2015 for the City of Bristol. Previously, he was Head of Industry and Waste Regulation at the Environment Agency, England and Wales.

Glossary of Terms

BAT	Best Available Techniques
BATNEEC	Best Available Techniques Not Entailing Excessive Cost
BERR	Department for Business, Enterpr*ise* and Regulatory Reform
BIS	Department for Business Innovation and Skills
BPEO	Best Practicable Environmental Option
BPM	Best Practicable Means
BRDO	Better Regulation Delivery Office
BRE	Better Regulation Executive
BREF	BAT Reference Document
BRU	Better Regulation Unit
CAP	Compliance Assessment Plan
CAR	Compliance Assessment Report
CCS	Compliance Classification System
CIWEM	Chartered Institution of Water and Environmental Management
CIWM	Chartered Institution of Wastes Management
COMAH	Control of Major Accidents and Hazards (Regulations)
DECC	Department for Energy and Climate Change

J. Foreman (ed.), *Developments in Environmental Regulation*, Palgrave Studies in Environmental Policy and Regulation, DOI 10.1007/978-3-319-61937-8

Defra	Department for the Environment, Food and Rural Affairs
DOENI	Department of the Environment, Northern Ireland
EA	Environment Agency
EAP	Environment Action Programme
EARA	Environmental Auditors Registration Association
EC	European Commission
EEA	European Economic Area or European Environment Agency
EEE	Electrical and Electronic Equipment
EGSS	Environmental Goods and Services Sector
EIB	European Investment Bank
EIPPCB	European IPPC Bureau (EIPPCB)
EMS	Environmental Management System
E-PRTR	European Pollutant Release and Transfer Register
EP	European Parliament
EPR	Environmental Permitting Regulations (England and Wales) 2010
ESA	Environmental Services Association
EU	European Union
GES	Good Ecological Status
HLG	High Level Group of Independent Stakeholders
HMIP	Her Majesty's Inspectorate of Pollution (1987 to 1996)
HMT	Her Majesty's Treasury
HSE	Health & Safety Executive
HWI	Hazardous Waste Inspectorate
IED	Industrial Emissions Directive
IEMA	Institute of Environmental Management & Assessment
IMPEL	European Union Network for the Implementation and Enforcement of Environmental Law
IPC	Integrated Pollution Control
IPPC	Integrated Pollution Prevention and Control
JRC	EU Joint Research Centre
LAPC	Local Air Pollution Control
LBRO	Local Better Regulation Office
MCERTS	Environment Agency Monitoring Certification Scheme
MPA	Mineral Products Association

NFRD	Non-Financial Reporting Directive
NPS	National Permitting Service (Environment Agency)
NRW	Natural Resources Wales
NSCA	National Society for Clean Air and Environmental Protection
OFWAT	Office for Water Services
ONS	Office for National Statistics
OPRA	Operational Risk Appraisal
PCB'S	Polychlorinated Biphenyls
POPs	Persistent Organic Pollutants
PPP	Pollution Pays Principle
RBMP	River Basin Management Plans
RCEP	Royal Commission on Environmental Pollution
REACH	Registration, Evaluation, Authorisation and Restriction of Chemicals
RPC	Regulatory Policy Committee
RRC	Reducing Regulation Cabinet Sub-committee
RTC	Red Tape Challenge
SEEA	System of Environmental Economic Accounting
SEPA	Scottish Environment Protection Agency
SERR	Smarter Environmental Regulation Review
UKELA	United Kingdom Environmental Law Association
UNECE	United Nations Economic Commission for Europe
USEPA	United States Environmental Protection Agency
UWWT	EU Urban Waste Water Treatment Directive
WEEE	Waste Electrical and Electronic Equipment
WFD	EU Water Framework Directive (WFD)

Index

J. Foreman (ed.), *Developments in Environmental Regulation*, Palgrave Studies in Environmental Policy and Regulation, DOI 10.1007/978-3-319-61937-8

F

G

Kolibri 360 GmbH
Ferdinand-Jühlke-Straße 7
99095 Erfurt, Deutschland
produktsicherheit@kolibri360.de